Christian Bermes

Design and dynamic modeling of autonomous coaxial micro helicopters

Christian Bermes

Design and dynamic modeling of autonomous coaxial micro helicopters

Towards palm-sized MAVs

Südwestdeutscher Verlag für Hochschulschriften

Impressum/Imprint (nur für Deutschland/only for Germany)
Bibliografische Information der Deutschen Nationalbibliothek: Die Deutsche Nationalbibliothek verzeichnet diese Publikation in der Deutschen Nationalbibliografie; detaillierte bibliografische Daten sind im Internet über http://dnb.d-nb.de abrufbar.
Alle in diesem Buch genannten Marken und Produktnamen unterliegen warenzeichen-, marken- oder patentrechtlichem Schutz bzw. sind Warenzeichen oder eingetragene Warenzeichen der jeweiligen Inhaber. Die Wiedergabe von Marken, Produktnamen, Gebrauchsnamen, Handelsnamen, Warenbezeichnungen u.s.w. in diesem Werk berechtigt auch ohne besondere Kennzeichnung nicht zu der Annahme, dass solche Namen im Sinne der Warenzeichen- und Markenschutzgesetzgebung als frei zu betrachten wären und daher von jedermann benutzt werden dürften.

Coverbild: www.ingimage.com

Verlag: Südwestdeutscher Verlag für Hochschulschriften GmbH & Co. KG
Heinrich-Böcking-Str. 6-8, 66121 Saarbrücken, Deutschland
Telefon +49 681 37 20 271-1, Telefax +49 681 37 20 271-0
Email: info@svh-verlag.de

Approved by: Zürich, Eidgenössische Technische Hochschule, Dissertation, 2010

Herstellung in Deutschland (siehe letzte Seite)
ISBN: 978-3-8381-2198-7

Imprint (only for USA, GB)
Bibliographic information published by the Deutsche Nationalbibliothek: The Deutsche Nationalbibliothek lists this publication in the Deutsche Nationalbibliografie; detailed bibliographic data are available in the Internet at http://dnb.d-nb.de.
Any brand names and product names mentioned in this book are subject to trademark, brand or patent protection and are trademarks or registered trademarks of their respective holders. The use of brand names, product names, common names, trade names, product descriptions etc. even without a particular marking in this works is in no way to be construed to mean that such names may be regarded as unrestricted in respect of trademark and brand protection legislation and could thus be used by anyone.

Cover image: www.ingimage.com

Publisher: Südwestdeutscher Verlag für Hochschulschriften GmbH & Co. KG
Heinrich-Böcking-Str. 6-8, 66121 Saarbrücken, Germany
Phone +49 681 37 20 271-1, Fax +49 681 37 20 271-0
Email: info@svh-verlag.de

Printed in the U.S.A.
Printed in the U.K. by (see last page)
ISBN: 978-3-8381-2198-7

Copyright © 2012 by the author and Südwestdeutscher Verlag für Hochschulschriften GmbH & Co. KG and licensors
All rights reserved. Saarbrücken 2012

Preface

This book is a reprint of my doctoral thesis DISS. ETH NO. 18847 that I successfully defended on February 26th, 2010 at the Swiss Federal Institute of Technology (ETH) Zürich. Supervisors were Prof. Dr. Roland Siegwart and Prof. Dr. Kenzo Nonami, defense chair was Prof. Dr. Robert Riener. It represents the research and thesis work I have performed at the Autonomous Systems Lab (ASL) from September 2006 to February 2010.
This work would not have been possible without the contributions and support of countless people. I would like to express my sincere gratitude to them.
It is my hope that the reader finds some interest in the theoretical and practical considerations of this book. If it gives some insight, that will be great. If it sparks off some new research ideas, that will be even better.

Solothurn, June 2012   Dr. Christian Bermes

Abstract

This thesis deals with the design, the dynamic modeling and the simulation of autonomous coaxial micro helicopters. The helicopter prototypes and modeling results that are presented are part of the project muFly, a European framework research project aiming at the design of an autonomous micro helicopter that is comparable to a small bird in size and mass.

The first goal of this work is to introduce two prototype designs for the muFly helicopter. The first prototype is designed in a modular way, to allow for fast sensor and actuator exchange and easy testing in different configurations. This modularity, however, has to be paid by quite high a total mass, which prevents from integration of the full set of sensors required for autonomous flight. The second prototype is designed in a highly integrated fashion to reduce the total mass and integrate the complete sensor set for autonomous flight, including the x-y-position sensor consisting of laser diodes and an omnidirectional camera. The key to achieve the reduced mass is the dual use of electronics as structure, which leads to a mass saving of approximately 17 %, while the sensor payload is increased.

The second goal of this work is to lay a more theoretical foundation for future coaxial micro helicopter designs, in particular focus are their ability to stabilize passively in roll and pitch, and their steering principles. Therefore, a modular dynamic model is developed, which incorporates the active and passive flapping characteristics of a hingeless rotor system, two optional steering principles, which are swash plate steering and steering by displacing the center of mass of the helicopter, and a stabilizer bar simulation module, which is also validated experimentally. Simulation results show the advantages of the coaxial rotor system in terms of a stronger disturbance rejection and a lower degree of total cross coupling. Moreover, swash plate steering is identified as the better steering option for coaxial micro helicopters, both in terms of output amplitude and energy consumption. Also the limitations of the stabilizer bar in terms of passive roll and pitch stabilization are shown.

Finally, the dynamic model is used for a design parameter study of the most relevant system design parameters. It aims at passive roll and pitch stabilization of the helicopter without a stabilizer bar. Simulation results show that passive stabilization can be achieved, and that the key parameter to passive stabilization is the rotor blade hinge offset, while blade flapping inertia and the distance between the rotors and the center of mass of the helicopter play a less important role.

Key words: Autonomous micro aerial vehicle, Coaxial helicopter, Dynamic simulation, Stabilizer bar, System design, System integration

Kurzfassung

Diese Arbeit behandelt den Prototypenentwurf, die dynamische Modellierung und die Simulation von autonomen koaxialen Mikrohelikoptern. Die gezeigten Prototypen und Ergebnisse sind Teil des Projektes muFly, einem europäischen Rahmenprojekt mit dem Ziel, einen autonomen Mikrohelikopter zu bauen, der in seiner Grösse und seinem Gewicht vergleichbar ist mit einem kleinen Vogel.

Ein Ziel dieser Arbeit ist die Einführung von zwei Prototypen des muFly-Helikopters. Der erste Prototyp ist als modulare Testplattform konzipiert, was den einfachen Austausch von Sensoren und Aktuatoren und Tests in unterschiedlichen Konfigurationen erlaubt. Der Nachteil dieser Modularität ist ein hohes Gesamtgewicht, so dass nicht alle für volle Autonomie notwendigen Sensoren integriert werden können. Der zweite Prototyp ist im Gegensatz dazu stark integriert, was zu einer Reduzierung der Gesamtmasse bei gleichzeitiger Mitnahme aller für autonomen Flug notwendigen Sensoren, insbesondere des aus Laserdioden und Omnidirektionalkamera bestehenden x-y-Positionssensors, führt. Die reduzierte Masse und der hohe Integrationsgrad werden durch duale Verwendung der Bordelektronik als Strukturelemente erreicht, so dass bei erhöhter Sensornutzlast die Gesamtmasse um etwa 17 % reduziert wird.

Ein weiteres Ziel dieser Arbeit ist es, eine stärker an der Helikoptertheorie orientierte Grundlage für zukünftige koaxiale Entwürfe zu schaffen. Von besonderem Interesse sind dabei die passive Roll- und Nickstabilität und die Auswahl eines geeigneten Steuerprinzips. Dazu wird ein modulares Modell für die dynamische Simulation eingeführt, welches die aktiv und passiv erregten Anteile der Schlagbewegung der Rotorblätter, zwei optionale Steuermechanismen in Form von Taumelscheibe und Massenschwerpunktverlagerung, sowie ein Modul für die Stabilisatorstange beinhaltet. Das simulierte Verhalten der Stabilisatorstange wird dabei auch experimentell validiert. Die Simulationsergebnisse zeigen die Vorteile des koaxialen Rotorsystems durch stärkere Robustheit gegenüber Störungen und einen geringeren Grad an Querkopplung der summierten longitudinalen und lateralen Schlagwinkel. Darüber hinaus kann durch quantitative Ergebnisse die Taumelscheibe als das bessere Steuerprinzip identifiziert werden, und die Grenzen der Stabilisatorstange für die passive Roll- und Nickstabilität aufgezeigt werden.

Schliesslich wird das dynamische Simulationsmodell für eine Studie der für die passive Roll- und Nickstabilität wichtigsten Entwurfsparameter genutzt. Ziel ist dabei, passive Stabilität ohne Stabilisatorstange zu erreichen. Die Ergebnisse der Studie zeigen, dass passive Stabilisierung durch richtige Auswahl von Entwurfsparametern erreicht werden kann, und dass der wichtigste Parameter dabei der Schlaggelenksabstand ist, während die Massenträgheit des Blattes und die Distanz zwischen den Rotoren und dem Helikoptermassenschwerpunkt eine untergeordnete Rolle spielen.

Stichworte: Autonomer Mikrohelikopter, Koaxialhelikopter, Dynamische Simulation, Stabilisatorstange, Systementwurf, Systemintegration

List of symbols

a_0		lift curve slope [1/rad]
a_1		longitudinal blade flapping angle [rad]
b		number of blades [-]
b_1		lateral blade flapping angle [rad]
c		chord length [m]
d		distance [m]
d		distance [m]
e		hinge offset [m]
g		gravitational acceleration [m/s^2]
h		shaft length [m]
k		rotational spring stiffness [Nm/rad]
k		aerodynamic coefficient [m]
l		length [m]
m		mass [kg]
r		radius [m]
v_1		inflow velocity [m/s]
A_1		lateral swash plate control input [rad]
B_1		longitudinal swash plate control input [rad]
D		normalized damping [-]
F		force [N]
I		moment of inertia [kgm^2]
L		Lagrangian function [kgm^2/s^2]
M		moment [Nm]
R		rotor radius [m]
T		thrust [N]
T		kinetic energy [kgm^2/s^2]
T_f		following time [s]
V		potential energy [kgm^2/s^2]
J		Jacobian matrix
M		mass matrix
D		damping matrix
C		spring matrix

α	stabilizer bar phase angle [°]
α_s	shaft angle of attack [rad]
β	blade flapping angle [rad]
ϵ_α	phase angle error [°]
η	longitudinal stabilizer bar flapping angle [rad]
γ	Lock number [-]
κ	gearing ratio [-]
μ	tip speed ratio [-]
ω	frequency [rad/s]
σ	rotor solidity [-]
θ_0	constant blade pitch [rad]
θ_1	blade twist [rad]
ξ	lateral stabilizer bar flapping angle [rad]
Ω	rotational speed [rad/s]
Ψ	rotor azimuth angle [rad]
$\vec{q}$	generalized coordinate vector
$\vec{r}$	position vector]
$\vec{v}$	translational velocity vector
$\vec{\lambda}$	Lagrangian multiplier vector
$\vec{\omega}$	rotational velocity vector
$\vec{\Gamma}$	search parameter vector

Subscripts and superscripts

aero	aerodynamic
bld	blade
cf	centrifugal
cm	center of mass
com	combined
drag	drag
eig	eigen-
eqv	equivalent
flap	flapping
fus	fuselage
fwd	forward
gen	generalized
gyro	gyroscopic
h	helicopter
imp	imprinted

in	inertial
lat	lateral
lift	lift
lo	lower rotor
long	longitudinal
max	maximal
mov	moveable
norm	normalized
rot	rotational
sb	stabilizer bar
sbm	stabilizer bar mass
sbr	stabilizer bar rod
sim	simulation
sp	swash plate
sprg	spring
trans	translational
up	upper rotor

Acronyms

ASL   Autonomous Systems Laboratory

BLDC   Brushless Direct Current

CSEM   Swiss Center for Electronics and Microtechnology

DC   Direct Current

ETHZ   Eidgenössische Technische Hochschule Zürich

FAI   Fédération Aéronautique Internationale

IMU   Inertial Measurement Unit

MAV   Micro Aerial Vehicle

MDO   Multidisciplinary Design Optimization

MOUT   Military Operations in Urban Terrain

RC   Radio Control

SAR   Search And Rescue

STFT   Short Time Fourier Transformation

STREP   Specific Targeted Research or Innovation Project

UAV   Unmanned Aerial Vehicle

Contents

Preface	i
Abstract	iii
Kurzfassung	v
List of symbols	vii
Acronyms	xi

1 Introduction   1
    1.1 Motivation and objectives 2
    1.2 State of the art 5
       1.2.1 Rotary wing UAV and MAV modeling and simulation 6
       1.2.2 Stabilizer bar modeling 7
       1.2.3 Design methodologies and optimization 9
       1.2.4 Rotary wing UAV and MAV prototype developments 10
    1.3 Contributions 12
    1.4 Structure of the thesis 13

2 Coaxial micro helicopter prototypes   15
    2.1 Prototype muFly 1 15
    2.2 Prototype muFly 2 19
    2.3 Prototype comparison 22
    2.4 Summary 26

3 Conventions and basic concepts   29
    3.1 Basic functionality of the coaxial helicopter 29
    3.2 Coordinate systems 30
    3.3 Rotor azimuth and blade degrees of freedom 32
    3.4 Rotor blade flapping 34
    3.5 Steering moments 37
    3.6 Helicopter equations of motion 41

| | | 3.7 Summary . . . . | 41 |

4 Stabilizer bar modeling and verification — 43
- 4.1 Physical system 43
- 4.2 Euler-Lagrange modeling of the system 45
  - 4.2.1 System outline and coordinates 46
  - 4.2.2 Kinetic and potential energy 49
  - 4.2.3 External forces and moments 52
  - 4.2.4 Stationary solution and linearization 53
  - 4.2.5 Determination of reaction moments 56
- 4.3 Simulation results 56
  - 4.3.1 Motion simulation 57
  - 4.3.2 Force simulation 58
- 4.4 Experimental setup 58
- 4.5 Signal processing 60
- 4.6 Experimental results 63
  - 4.6.1 Following time experiments 63
  - 4.6.2 Phase angle experiments 67
- 4.7 Summary 69

5 Modular dynamic model — 71
- 5.1 Model structure 71
- 5.2 Hingeless rotor module 75
  - 5.2.1 Basic model and parameter equivalence 75
  - 5.2.2 Derivation of the rotor forces and moments 76
  - 5.2.3 Derivation of the blade flapping angles 77
- 5.3 Stabilizer bar module 85
- 5.4 Swash plate steering module 87
- 5.5 Center of mass displacement steering module 88
- 5.6 Simulation results 92
  - 5.6.1 Stabilizer bar 92
  - 5.6.2 Swash plate steering 95
  - 5.6.3 Center of mass displacement steering 98
  - 5.6.4 Comparison 98
- 5.7 Summary 104

6 Parameter study for passive roll and pitch stability — 107
- 6.1 Basic model and parameter selection 107
- 6.2 Software implementation 110
- 6.3 Parameter study results 112

	6.3.1	Stabilizer bar considerations	112
	6.3.2	Rotor distance and combined hinge offset	114
	6.3.3	Combined hinge offset and blade flapping inertia	116
	6.3.4	Rotor distance and blade flapping inertia	116
	6.3.5	Parameter selection	121
6.4	Summary	123

7 Conclusion — 125

A Historical evolution of coaxial helicopters — 129

B Matrices for the stabilizer bar model — 135
  B.1 Transformation matrices 135
  B.2 Linearized system matrices 136
      B.2.1 Mass matrix $\mathbf{M}(t)$ 136
      B.2.2 Damping matrix $\mathbf{D}(t)$ 137
      B.2.3 Spring matrix $\mathbf{C}(t)$ 138

Bibliography — 139

List of Tables

2.1	Mass distribution of the prototype muFly 1.	18
2.2	Mass distribution of the prototype muFly 2.	23
3.1	Baseline parameters for the steering moment variation.	39
4.1	Parametrization of the stabilizer bar model.	54
4.2	Simulated stabilizer bar following times.	58
4.3	Experimental stabilizer bar following times.	66
5.1	Parameters of the steering principle comparison.	102
6.1	Design parameter ranges for the variations.	112
6.2	Stabilizer bar parameter ranges for the variations.	112

List of Figures

1.1	Large and populated indoor areas for surveillance.	2
1.2	Typical disaster scenes for indoor helicopter search.	2
1.3	Possible application areas for plant supervision.	3
1.4	Mine and cave environment.	3
1.5	Window and door breaching.	4
1.6	An artist's vision of muFly.	5
1.7	The Proxflyer stabilization system and the stabilizer bar.	8
1.8	The micro helicopters μFR and MICOR.	11
2.1	Assembled prototype muFly 1.	16
2.2	Xsens MTi OEM IMU for the prototype muFly 1.	16
2.3	Central frame as CAD model.	17
2.4	Assembled prototype muFly 2.	20
2.5	Disassembled prototype muFly 2.	21
2.6	Position sensors of the prototype muFly 2.	22
2.7	Mass distribution of the prototype muFly 1.	24
2.8	Mass distribution of the prototype muFly 2.	24
2.9	Total mass comparison between the prototypes.	25
2.10	Functional group mass comparison between the prototypes.	26
2.11	Prototypes muFly 1 (left) and muFly 2 (right) in flight.	26
3.1	Yaw and vertical motion of the coaxial micro helicopter.	30
3.2	Roll and pitch motion of the coaxial micro helicopter.	31
3.3	Inertial and body-fixed coordinate frames $\{I\}$ and $\{B\}$.	31
3.4	Sign conventions for the upper and lower rotor.	33
3.5	Degrees of freedom of an articulated rotor head.	33
3.6	Schematics of the hingeless rotor and its analogous model.	34
3.7	Longitudinal and lateral flapping angle conventions.	36
3.8	Phase delay of an oscillatory system.	37
3.9	Two types of steering moments on the hingeless coaxial rotor.	38
3.10	Variation of the steering moments.	40
4.1	The stabilizer bar in schematics and reality.	44

4.2	Definition of the stabilizer bar phase angle α.	44
4.3	Different variations of the Bell stabilizer bar.	45
4.4	Schematic view of the multi-body system.	47
4.5	Motion simulation result for the stabilizer bar flapping angle.	57
4.6	Force simulation result for the stabilizer bar phase angle.	59
4.7	Schematic of the stabilizer bar test rig.	60
4.8	The test rig at the ASL.	61
4.9	Unfiltered measurement of the moment M_y.	61
4.10	STFT of the example signal.	62
4.11	STFT of a measurement with no motors running.	63
4.12	Test rig tilt for the following time experiments.	64
4.13	Moment trajectories and following times for four different stabilizer bar inertias at a rotor speed of 270 rad/s.	64
4.14	Moment trajectories and following times for four different stabilizer bar inertias at a rotor speed of 300 rad/s.	65
4.15	Moment trajectories and following times for four different stabilizer bar inertias at a rotor speed of 330 rad/s.	65
4.16	Moment trajectories and following times for four different stabilizer bar inertias at a rotor speed of 360 rad/s.	66
4.17	Comparison between simulation and experiments at different rotor speeds.	67
4.18	Test rig tilt for the phase angle experiments.	68
4.19	Moments M_x and M_y as functions of the phase angle.	68
5.1	Model structure for swash plate steering.	72
5.2	Model structure for center of mass displacement steering.	73
5.3	Simulink implementation of the modular simulation model.	74
5.4	Equivalence of hinge offset and flapping spring stiffness.	75
5.5	Projection of the rotor thrust T on the body-fixed axes.	76
5.6	Rotor forces and moments for a roll moment disturbance of 1 Nmm without hinge offset.	80
5.7	Rotor forces and moments for a roll moment disturbance of 1 Nmm with hinge offset 10 mm.	80
5.8	Rotor forces and moments for a pitch moment disturbance of 1 Nmm without hinge offset.	81
5.9	Rotor forces and moments for a pitch moment disturbance of 1 Nmm with hinge offset 10 mm.	81
5.10	Rotor forces and moments for an x-force disturbance of 10 mN without hinge offset.	84
5.11	Rotor forces and moments for an x-force disturbance of 10 mN with hinge offset 10 mm.	85

5.12	Rotor forces and moments for a y-force disturbance of 10 mN without hinge offset.	86
5.13	Rotor forces and moments for a y-force disturbance of 10 mN with hinge offset 10 mm.	86
5.14	Schematic of the center of mass steering mechanism.	88
5.15	Comparison of the exact and simplified CM steering model.	91
5.16	Inverse response behavior of the exact model.	92
5.17	Trajectories for a roll moment disturbance of 30 Nmm with hinge offset 5 mm and different stabilizer parametrizations.	93
5.18	Rotor forces and moments for a roll disturbance of 30 Nmm with hinge offset 5 mm and stabilizer following time 250 ms.	94
5.19	Rotor forces and moments for a roll disturbance of 30 Nmm with hinge offset 5 mm and stabilizer following time 500 ms.	95
5.20	Trajectories for a roll steering input of 5° with different hinge offsets and without stabilizer bar.	96
5.21	Rotor forces and moments for a roll steering input of 5° without hinge offset and without stabilizer bar.	97
5.22	Rotor forces and moments for a roll steering input of 5° with hinge offset 10 mm and without stabilizer bar.	97
5.23	Trajectories for a roll steering input of 5° with different hinge offsets and with stabilizer bar.	99
5.24	Rotor forces and moments for a roll steering input of 5° without hinge offset and with stabilizer bar.	100
5.25	Rotor forces and moments for a roll steering input of 5° with hinge offset 10 mm and with stabilizer bar.	100
5.26	Trajectories for a CM roll steering input of 10 mm to the simplified CM steering model with different hinge offsets.	101
5.27	Trajectories for swash plate and CM steering.	103
6.1	Model structure for the parameter search.	108
6.2	Roll disturbance moment for the parameter search.	109
6.3	Flow chart of the parameter search.	111
6.4	Variation of phase angle error and following time.	113
6.5	Variation of phase angle error and following time (zoomed).	113
6.6	Variation of rotor distance and combined hinge offset (1).	114
6.7	Variation of rotor distance and combined hinge offset (2).	115
6.8	Variation of rotor distance and combined hinge offset (3).	115
6.9	Variation of rotor distance and combined hinge offset (4).	116
6.10	Variation of combined hinge offset and flapping inertia (1).	117
6.11	Variation of combined hinge offset and flapping inertia (2).	117

6.12 Variation of combined hinge offset and flapping inertia (3). 118
6.13 Variation of combined hinge offset and flapping inertia (4). 118
6.14 Variation of rotor distance and flapping inertia (1). 119
6.15 Variation of rotor distance and flapping inertia (2). 119
6.16 Variation of rotor distance and flapping inertia (3). 120
6.17 Variation of rotor distance and flapping inertia (4). 120
6.18 Final variation of rotor distance and flapping inertia (1). 121
6.19 Final variation of rotor distance and flapping inertia (2). 122
6.20 Final variation of rotor distance and flapping inertia (3). 122
6.21 Final variation of rotor distance and flapping inertia (4). 123

A.1 Lomonosov's Aerodynamic, a toy by Launoy and Bienvenu, and Henry Bright's patent. 129
A.2 The Breguet-Richet Gyroplane No.1. 130
A.3 Sikorsky's S-1 and Antonov's Helicoplane. 130
A.4 The Pescara No.3 and D'Asciano's record helicopter. 131
A.5 The Breguet G.IIE Gyroplane and Manzolini's Libullella. 131
A.6 Utility helicopter Ka-32 and combat helicopter Ka-50. 132
A.7 Coaxial helicopter UAVs: Gyrodyne QH-50 and Ka-37. 132
A.8 Gyrodyne prototype GCA2 and future design Sikorsky X2. 133
A.9 Timeline for the early days of the coaxial helicopter. 133
A.10 Timeline for the modern ages of the coaxial helicopter. 134

Chapter 1

Introduction

The history of powered Unmanned Aerial Vehicles (UAVs) is almost as long as powered manned aviation itself. With 'war as the father of all things', the first intended practical purpose of powered UAVs was to carry ammunition to a target without risking the life of a pilot. In 1918, this was achieved for the first time with the fixed-wing Curtiss/Sperry Aerial Torpedo [67], giving the signal to start almost a century of powered unmanned flight. Until the first unmanned helicopters went to flight, it took a little longer, mostly because the much more complex helicopter technology had to be mastered first. One of the first purposeful unmanned helicopters was the Gyrodine QH-50 in 1961. It had been developed for ship-based operations, was extensively tested [72], and saw field use for many years. The race for helicopter UAVs of all kinds was now open, however, still mostly driven by the military [33, 36].
With the availability of small and lightweight actuators and energy sources, and inseminated by the helicopter modeler community [86] and toy industry [28], a completely new era of unmanned helicopter flight began at the end of the last century [55, 96]. Suddenly, the lower size limit was shifted to rotor diameters of 15 cm and smaller, and besides the rather classic rotor configurations (conventional main and tail, coaxial, tandem and quadrotor), the most exotic options for propulsion and steering have been realized, may it be for practical purposes or out of mere curiosity. These developments include shrouded quadrotors [73], ducted fans [54], cyclocopters [15] or biologically inspired samara vehicles [90, 97], which are in essence one bladed rotors. Even the use of insects as a power source for small aerial vehicles has been investigated [78]. A new class of aerial vehicle has been born – the Micro Aerial Vehicle (MAV).
With so much freedom in the design of systems, and with the relatively cheap availability of components to build such systems, MAVs have grown to a scientific research area of their own, no longer restricted to military needs and funding. This thesis aims to contribute to the research area in the domain of coaxial rotary wing MAVs that can operate fully autonomously in indoor environments.

Figure 1.1: Large and populated indoor areas for surveillance (Source: http://muc.in/).

Figure 1.2: Typical disaster scenes for indoor helicopter search as part of an SAR mission (Source: http://www.adrc.asia/).

1.1 Motivation and objectives

There are many scenarios, for which the use of an autonomous helicopter in an indoor environment is desirable. All of these scenarios have in common that they offer an unstructured environment, which is often difficult to traverse or inaccessible for ground vehicles, and which requires the ability of a vehicle to maneuver in the vertical direction and to hover. The most likely scenarios are:

- **Surveillance** of buildings and large indoor areas, such as airports, train stations or exhibition areas (see Figure 1.1). For security reasons, these areas are currently kept under surveillance by fixed cameras, which produce a lot of data to be reviewed, or by security personnel, which is very cost intensive. An autonomous helicopter performing this task could reduce the amount of data to be processed, since it scans only very selectively, while it is cheaper and probably deployed much faster than security personnel.
- **Search** and Rescue (SAR) missions, which require a reconnaissance phase in areas that are difficult or dangerous to access for rescue personnel and not traversable for ground vehicles due to rubble on the ground or collapsed staircases (see Figure 1.2). Equipped with a range sensor, the helicopter could map the environment to support the situation assessment and identify feasible access routes for rescue personnel. Equipped with an infrared camera, the helicopter could detect victims and transmit their position to a ground station to facilitate a targeted rescue attempt. For the search phase of such SAR missions, indoor helicopters can be a very valuable tool.

 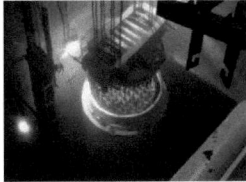

Figure 1.3: Possible application areas for plant supervision in regular (Source: http://www.oregonlive.com/) and hazardous environments (Source: http://www.par.com).

Figure 1.4: Partially collapsed mine (Source: http://www.flickr.com) and cave environment (Source: http://photography.nationalgeographic.com/) for a mapping scenario.

- **Plant surveillance and inspection**, especially in hazardous areas of chemical or nuclear plants (see Figure 1.3). For mobile supervision of plant processes, the helicopter could carry a visual camera, gas or radiation sensors. Operation of helicopters in these environments could reduce plant down time to save costs, since the supervision or inspection could be executed while the plant is running. Moreover, in accident scenarios of chemical or nuclear plants, indoor helicopters could be used for prescreening of the sites, before personnel would have to be put at risk.
- **Exploration** and mapping of mines and caves, which is important for the prevention of subterraneous collapses (see Figure 1.4) and the protection of surface structures from subsidence. Moreover, cave mapping is very interesting from a scientific point of view. The task is difficult to perform with ground vehicles in the face of rock or water obstacles, or if vertical steps are too high. A swarm of indoor helicopters could set up a communication relay and transmit mapping data collected with on board range scanners.
- **Police and military** operations, which involve entering of buildings and breaching of rooms as in hostage rescue scenarios and Military Operations in Urban Terrain (MOUT) (see Figure 1.5). Indoor helicopters could be used to enter a room first and assess the situation, before operators need to put their lives at risk. With the indoor helicopter, room penetration could become much more flexible, since in addition to doors, any other opening of the room (windows, wall breaches) could be used.

To master all of these complex scenarios, the indoor helicopter has to be fully autonomous, i.e. able to operate independently of a ground station if necessary. For full autonomy,

Figure 1.5: Window breaching by the police (Source: dpa (German Press Agency)) and door breaching by military forces (Source: http://www.spiegel.de).

this helicopter has to carry all sensors required and a sufficient amount of computational power to allow for onboard automatic control. Furthermore, in order to be able to traverse small openings, the helicopter must be very compact to be of practical use. Finally, full autonomy also means energy autonomy. To be mission capable, the helicopter must have a flight endurance of at least ten minutes. Summarizing these requirements, the autonomous indoor helicopter has to be small and compact, highly integrated, energy efficient and optimized in mass wherever possible. A new class of UAV needs to be developed – the fully autonomous helicopter MAV.

In order to provide a solution to this need, a European consortium of six partners both from academia and industry assembled and started the project muFly in June 2006. Official aim of the project was to design, build and test a fully autonomous micro helicopter, which should be comparable to a small bird in mass and size [8]. For the project duration of 36 months, muFly was funded as a Specific Targeted Research or Innovation Project (STREP) under the Sixth Framework Programme of the European Commission. The vision was to provide a system that could function in the previously mentioned scenarios, just as the artist's vision in Figure 1.6 shows: a rescue team deploys several muFly helicopters to reconnoiter a building in danger of imminent collapse. Within the project, the research and development tasks were distributed according to the respective field of expertise of the partners. Those were:

- Albert-Ludwigs-Universität Freiburg, Department of Computer Science [1]: navigation and localization algorithms research [47].
- Eidgenössische Technische Hochschule Zürich (ETHZ), Autonomous Systems Laboratory (ASL) [2]: system integration, design optimization, controller design.
- Technische Universität Berlin, Berlin Center of Advanced Packaging [3]: battery and micro fuel cell research.
- CEDRAT Groupe [4]: motors and actuators.
- Swiss Center for Electronics and Microtechnology (CSEM) [5]: x-y-position detection using an omnidirectional camera [40].
- Xsens Technologies B.V. [11]: Inertial Measurement Unit (IMU) development.

A contribution to the work packages of ETHZ is this thesis. Motivated by the various applications for a system like the muFly MAV, the objective is to propose and realize con-

Figure 1.6: An artist's vision of muFly helicopters being deployed by a rescue team in a disaster scenario (Source: http://www.mufly.org).

crete designs for the project helicopter and to provide dynamic models for coaxial micro helicopters in differing setups, to support design decisions based on qualitative and quantitative comparison of these models.

Nowadays, research and toy helicopter hingeless rotor systems are mostly experience-based designs, proving themselves flightworthy only after a tedious trial and error process. Once the state of flightworthiness is reached, it is oftentimes not exactly clear, which modification of the system made the difference and why. The objective of this thesis is to give future designs a more theoretic foundation and to explain, which design parameters are important and which ones are not for coaxial helicopters on the MAV scale. A special objective within this work are considerations about the passive roll and pitch stability of coaxial micro helicopters, either reached by use of an additional device, the stabilizer bar [27], or by advantageous selection of the rotor blade properties of the rigid rotor system [17]. Finally, design propositions and their mechanical realizations up to the level of test flight exemplify the challenge of system integration for the various subsystems of an autonomous micro helicopter.

The issues of aerodynamic rotor blade optimization for coaxial micro helicopters and attitude and position control are not within the scope of this thesis. Detailed research on both topics can be found in [82].

1.2 State of the art

Research and development in the field of unmanned mini and micro helicopters has been very active especially in the past ten years, when these systems became the focus of a rapidly growing number of research groups. In this section, the state of the art is considered for the domains of research that are covered by this thesis. These are the general field of rotary wing UAV and MAV modeling, the specific field of stabilizer bar modeling, design methodologies and design optimization applied to UAVs and MAVs, and descriptions of actual working prototypes of these systems. In the following, an overview over the literature in these fields of research is given.

1.2.1 Rotary wing UAV and MAV modeling and simulation

In general, there exists a very large number of simulation models and results for rotary wing Unmanned Aerial Vehicle (UAV)s and MAVs. It seems that almost every researcher that needs a simulation model, creates his own, regardless of the availability of existing models. The literature review in this section is restricted to rotor configurations that are relevant to the research topic of this thesis, i.e. coaxial helicopters. Therefore, only models for conventional and coaxial helicopters are considered. Quadrotors and other 'exotic' rotor configurations are excluded from the simulation model review.

For small helicopters in conventional rotor configuration, that is with a single main and tail rotor, one of the standard and most cited references is [63]. This model reflects all the important mechanic and aerodynamic subsystems of the helicopter and is subsequently used for system identification with test flight data. A similar procedure is followed in [48]. The proposed models show good correlation with real test data and are very useful for controller synthesis. The only difficulty is that they rely on the existing hardware and oftentimes incorporate artificial parameters, such as time constants, which do not represent direct design parameters of the helicopter. Therefore, analysis of the flight dynamics of an envisioned, not yet built prototype is difficult with such models.

Further work in this domain is shown in [92], [91] and [12], with all of them focussing on modeling for controller and autopilot design. While all of the previously mentioned resources focus on forward simulation, i.e. the response of the system to a given input, inverse simulation is investigated in [12], which is quite uncommon in the field of helicopter simulation. Here, the desired output trajectories are generated first. Subsequently, they are fed into the dynamic system to find the input trajectories that lead to the desired output. The difficult point, however, is the system inversion, which is not always possible. Research on the topic of hingeless rotors is significantly younger than helicopter research in general. This is due to the fact that hingeless rotors became only available with increased sophistication of blade materials, and due to the fact that the large distribution of able toy and research mini and micro helicopters is a phenomenon of the recent years. Literature relevant for this thesis are [58] and [56], which both analyze the effects of the hingeless rotor design on blade flapping motion and helicopter flight dynamics. The first does this for a full scale helicopter, the latter analyzes a mini helicopter system, however, both in a conventional rotor setup.

The review of conventional configuration simulation models is concluded by [14] and [41]. While specialized aeroelastic simulation software and very complex analysis are used to examine helicopter flight stability in [14], in [41] a modular and highly adaptable helicopter model has been implemented in Matlab/Simulink, which allows for analysis of various subcomponents while remaining in the standard Simulink environment. This facilitates subsequent optimization or controller synthesis. For practical purposes [41] is certainly one of the best simulation models currently available.

The only published resource on complete modeling of full scale coaxial helicopters is a Russian work [13]. Presented in a very abbreviated notation and with only sparse explanations, the work is of little practical use and only listed here for completeness.

For small coaxial helicopters, the number of sophisticated simulation models is smaller than it might be expected. On the lower end of complexity are [29] and [38], which introduce very simple models without rotor head dynamics. While these models can be sufficient for controller design, they are certainly not for design analysis and optimization, which require more detailed descriptions of the rotor head and stabilizer bar dynamics.

Models with a higher degree of complexity can be found in [52] for a coaxial helicopter with stabilizer bar dynamics, and in [93] for a novel design that combines the ducted fan principle with the coaxial rotor configuration. The model in [52], however, does also not seem to include rotor head dynamics. Furthermore, the work is available in Japanese only, and hence of little practical use. Another work for a coaxial ducted fan is [54]. A very good coaxial model for controller design is [83], which incorporates the most important physical effects and shows very good correlation with flight data after parameter identification. The limitation is again that especially for rotor and stabilizer bar the design parameters are hidden in time constants, making model based helicopter design difficult.

An interesting approach to UAV modeling is taken in [31]. In that work, an attempt is made to develop a unified, generic model to cover several UAV types, specifically the fixed wing plane, the quadrotor, the ducted fan and the conventional configuration helicopter. While this is an interesting approach for simplified control models, it is not suitable for design simulation, where certain characteristics specific to the type of UAV need to be taken into account.

In summary, the most relevant works are either highly specialized models for conventional configuration helicopters, or practical models for small coaxial helicopters, which are mainly aimed at controller design. Models that allow for studying the influence of physical design parameters, especially those of the rotor system, have not been found.

1.2.2 Stabilizer bar modeling

On the specific issue of passive helicopter stabilization and dynamic modeling of the stabilizer bar, there are resources of varying level of detail and quality available.

For the topic of passive roll and pitch stabilization without a stabilizer bar, the significant

Figure 1.7: The Proxflyer stabilization system [66], and the stabilizer bar on the upper rotor of a coaxial helicopter.

contribution is the patent of the Proxflyer rotor system [65]. The basic principle involves metal rings attached to the blade tips of each rotor (see Figure 1.7) and thus additional inertia to create a slower following of the rotor discs with respect to fuselage roll and pitch motions. The system is successfully in use on simple toy helicopters and is described in a little more detail in [66], however, only qualitatively. Current shortcomings of this solution are that it has never been seen in combination with a swash plate steering mechanism, and that there is no formal quantitative description of its behavior, which makes a priori sizing very difficult.

Several researchers have published their work about the influence of a stabilizer bar on a helicopter. For the most part, these works are focused on single rotor configurations. Notable contributions are [74], [20] and [63]. Their modeling is mostly driven by the need for a suitable control model, which requires simplicity. Therefore, the authors have introduced time constants to describe the delayed following of the stabilizer after a fuselage roll or pitch motion. While this is correct from a technical point of view and advantageous for the control engineer, these following times need to be identified experimentally and cannot be predicted in the design phase of a helicopter, since their models lack the degree of detail for that (design parameters are not directly incorporated).

Works that go into more detail concerning the actual physics are the thesis works [34, 35]. There, the modeling is done for a conventional rotor, and the coupling of an aerodynamic fly bar into the swash plate steering inputs (Bell-Hiller system) is analyzed. However, considerations concerning design parameter selection for this system are not made. In a similar direction points [87], where it is proposed to include the stabilizer bar in the pitching inertia of the blades. While this assumption is certainly correct, it neglects to consider the function of the stabilizer bar as a feedback loop that sets the cyclic blade pitch inputs according to the roll and pitch velocities of the helicopter. Furthermore, the stabilizer bar is considered in textbook literature, for instance in [27]. Here, however, only the phase angle of the stabilizer bar with respect to the blade pitching axis is considered, while its following dynamics are left out. For the phase angle, the analysis is performed for a teetering rotor system, which always leads to a phase of 90° between stabilizer and blades. Analysis of the stabilizer in conjunction with a hingeless rotor system, which raises the question of phase angle tuning, is not given.

Finally, one reference is available that aims at the specific problem of correctly tuning a stabilizer bar for a hingeless rotor coaxial toy helicopter [88]. This work involves theoretical analysis and practical results, which match very well. It is, however, not clear, why different phase angles are found for identical rotors. Since it is not obviously comprehensible why a steering input from a stabilizer bar should be different from that of a swash plate mechanism in terms of the phase angle, the analysis given in the reference might not be completely correct.

The idea of mounting a rotor system on a test rig to measure rotor forces and moments effected by helicopter flight maneuvers is initially reported in [39] for small helicopters. Here, a test rig is used to measure reaction forces and torques to a linear displacement of the helicopter. The work is of high quality, experimental results are used to calculate stability derivatives for the system. The only restriction is that the test rig can only be used for linear and not for roll and pitch motions of the rotor system.

In summary, there is a broad variety of sources available on the stabilizer bar, but only a small portion of them is suitable for the special domain of micro helicopters. Especially for the stabilizer bar in connection with hingeless rotor systems, there is a need for better theoretical analysis and experimental validation of both its following and phase angle characteristics.

1.2.3 Design methodologies and optimization

Despite the large number of models available for rotary wing MAVs, only few authors have used their models to aim at stringent design methodologies or the optimization of their designs. Their work is recapitulated in the following.

As in the previous sections, most works on this topic can be found for helicopters in the conventional configuration. In terms of size, these works are all aimed at full scale manned helicopters, in terms of scope, they are usually aimed at the rotor system alone. In [30], extensive analysis of blade flapping on a single rotor is shown. The influence of different design parameters on system outputs such as flapping amplitude and rotor disc following time is examined. The results, however, are not integrated into a full helicopter simulation model to study their influence on the flight dynamics. Nevertheless, it is a standard reference on rotor analysis.

Other works, which aim at Multidisciplinary Design Optimization (MDO), are [42] and [43]. There, MDO is applied to the main rotor with the goal to optimize aerodynamic and mechanic properties of a rotor blade with respect to blade lagging. While certainly relevant for full scale helicopters, blade lagging can be neglected for micro helicopters. Research that covers the general sizing of full scale conventional helicopters is presented in [79]. Sizing laws are derived using system parameters of a large number of helicopter types and statistical methods. This approach is certainly interesting, however, it relies on the availability of large data quantities, certain payload assumptions, and cannot be transferred to other configurations like coaxial or tandem helicopters without the need for a new large data set to perform statistical methods.

The only works aiming at conventional helicopters on a smaller scale are [64] and [89]. In the first work, scaling and its effect on the helicopter stability derivatives are considered. A comparison is performed between a full scale and a small-scale helicopter with a rotor diameter of approximately three meters. The second work gives design guidelines for rotary wing MAVs under the aspect of low Reynold's number aerodynamics and focuses on the aerodynamic rotor blade design.

Considering the domain of coaxial micro helicopters, the foundation of published work becomes rather weak. The only noteworthy works are [22, 23], which are mainly concerned with aerodynamic and propulsion group optimization.

A significantly larger number of research works can be found on the design of quadrotors, most likely because they are currently the most popular research platforms due to their high payload capacity, simple mechanics and agile maneuvering options. Design guidelines with respect to dynamic behavior can be found for instance in [75, 76]. With respect to optimal propulsion component selection to maximize flight time, [24, 26] are the relevant references. Finally, in terms of placing system components to achieve a defined position of the center of mass or to meet certain requirements for the mass moments of inertia, a series of works has been published [68–70]. Here, different optimization strategies such as genetic algorithms have been successfully applied.

Another, rather exotic source is [51], with a cyclocopter simulation model used for design optimization. While the actual physical system is quite uncommon, the strategy to set up a model for design optimization purposes is certainly an inspiration for the dynamic modeling done in this thesis.

Summing up the literature for design methodologies and optimization, it can be seen that for coaxial micro helicopters, there is still some potential to use dynamic simulation models for design purposes.

1.2.4 Rotary wing UAV and MAV prototype developments

Eventually, all research results cumulate in the development of an actual prototype. Especially for helicopters, this can be hard work and the source of numerous problems, mostly related to the rotor vibrations. Therefore, successful helicopter prototyping always requires a strong base of crafting experience, as it can be usually found in the modeler community. It is therefore not a surprise that some of the best resources on the actual making of small-scale helicopters, even though remotely controlled and not autonomous, have been written by the 'gurus' of the modeler community. One of them is Dieter Schlüter [85, 86], who is known as the father of the Radio Control (RC) helicopter. In his work, he gives detailed procedures on sizing, building and tuning small helicopters, and especially their

 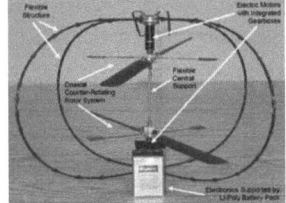

Figure 1.8: The µFR by Chiba University and Epson [94], and a MICOR prototype by the University of Maryland [81].

rotor systems. Of similar content and quality is the work of Bernet [19], another German RC helicopter enthusiast. Both resources are very valuable for beginners and newcomers to the topic, and are extremely helpful in order to avoid common mistakes. On the downside, these works are purely experience based, only applicable to helicopters in the conventional rotor configuration, and are not faced with the integration challenges that the designer of an autonomous micro helicopter has to cope with. The main focus of these works is in a rotor diameter range between 1 m and 3 m.

Apart from the modeler community, there are several works in the robotics domain that report on actual prototype building and which handle component integration and sizing challenges. In [25], the integration of a center of mass displacement steering mechanism, attitude sensors and a sensor for the ground clearance on coaxial helicopter with a rotor diameter of 35 cm is shown. Using the same steering principle, in [94] the smallest robotic helicopter known to date [7] is described. While the massive challenges in mechanical design and actuator integration for a helicopter with a mass of 12.3 g and a rotor diameter of 13.6 cm are solved superbly, the question of on board sensing and computing remains open. Attitude and position control signals are computed on a ground station computer and sent to the helicopter via an RC link, and the pose and position estimates are delivered by a ground-fixed camera system.

Some of the more exotic prototypes are the ones shown in [81] for steering using a smart structure airframe for the helicopter, in [22, 23] for a coaxial rotor MAV and in [93] for a coaxial rotor ducted fan. While being interesting for steering principle evaluation and simulation validation, these prototypes do not leave the level of RC models in terms of sensing and on board computing power. Despite this lacking degree of integration, they remain successful proofs of concept.

A very interesting prototype is shown in [62], where the development of the competition helicopter MAVSTAR is shown. This helicopter is based on a retrofitted toy helicopter and features an IMU, ultrasonic range finder and video camera. The gathered sensor data, however is processed on a ground station and respective control signals are sent back to the helicopter. Also, with a span of 30 cm and a mass of almost 500 g, the helicopter is larger than the previously mentioned prototypes.

Remarkably, all of these works feature coaxial rotor systems with various steering mechanisms. Apart from quadrotors, the coaxial rotor configuration is the configuration of choice for most successful robotic MAVs.

1.3 Contributions

This thesis aims at establishing a solid theoretical foundation for future designs of coaxial mini and micro helicopters. The focus is clearly on theoretical and simulative work, with some experimental data provided to support the simulated results. The models through the course of this thesis are derived from first principles. Therefore, they are based on meaningful physical parameters instead of artificial or lumped parameters. The latter have advantages in identification models used for controller synthesis, where lean models are needed to allow for model based control and implementation of the controllers on embedded systems with limited computational power. But if the modeling is intended to support design and design optimization, a high level of detail in the model is not a problem, since computational resources are available. This makes it possible to have direct influence on the model parameters through the design parameters, without the indirection that an artificial parameter introduces.

In summary, the contributions of this thesis are the following:

- Firstly, the thesis proposes two prototype designs for the project helicopter muFly. These designs show the development towards a maximal integration of the system components, while the total mass of the system is significantly reduced. However, these prototypes also show the current research gap in terms of simulation models that can be used for prototype evaluations before the actual design is realized.

- An analytical model of the stabilizer bar, which is a stability augmentation device used on almost every coaxial MAV that is currently in service, is derived. Together with a model derivation based on the Euler-Lagrange methodology, and corresponding dynamic simulation results for the following behavior and phase angle, experimental results from test rig measurements are shown and used to verify the model in terms of following time and phase angle.

- A modular simulation model for the design of coaxial mini and micro helicopters with hingeless rotors is developed. The model is modular by allowing to simulate a fully passive helicopter without steering inputs, a swash plate steered helicopter, and a helicopter that is steered by displacing its center of mass. Furthermore, an optional stabilizer bar module is included. It is therefore possible to observe the possibility of passive stabilization by the rotor system alone or augmented by a stabilizer bar, and to simulate and compare the performance of the two most likely steering principles on a coaxial rotor MAV.

- Based on the dynamic simulation model, a software structure for parameter variation is introduced. It allows for studying the influence of the stabilizer bar design parameters on the dynamic behavior of the helicopter, and, more importantly, leads to design parameters for a helicopter that is passively stable in roll and pitch without the use of a stabilizer bar.

This leads directly to the structure of the thesis in the following section.

1.4 Structure of the thesis

After this introduction explaining the motivation for this work, the state of the art and the main contributions, the two muFly prototypes are shown and compared in Chapter 2. For the further theoretical considerations, the basic definitions, conventions and some basic concepts of helicopter theory, which will be frequently used in the course of this work, are given in Chapter 3.

In Chapter 4, at first the functional principle of the stabilizer bar as a stability augmentation device is explained. This introduction into the system is followed by an analytical model derived using the Euler-Lagrange formalism, in order to obtain equations of motion for the coupled system of stabilizer bar and rotor blades. Based on this model and on a parametrization that corresponds to the rotor system of muFly, the following time and the optimal phase angle for the stabilizer bar are found in simulation. Subsequently, the experimental validation of the theoretical results is performed. A test rig for measuring rotor forces and moments is introduced, and experimental data with the respective signal processing is shown to find the following time and the optimal phase angle of the stabilizer bar experimentally.

In Chapter 5, the modular simulation model for coaxial micro helicopters is derived. It features the two steering options that are mostly seen on these systems: either swash plate steering or steering by displacing the center of mass of the helicopter. The modular architecture of the simulation model is explained, the underlying equations of each module are derived, and simulation results for several options of the model are given.

The core of this simulation model, which is the hingeless rotor model, is then used in Chapter 6 to study the dynamic behavior of a helicopter without steering inputs and without the stabilizer bar, which is desired to have passive roll and pitch stability solely by advantageous selection of certain design parameters of the rotor system.

While each chapter already contains a short and specific chapter summary, the thesis itself is concluded in Chapter 7. Also, a brief outlook on possible future designs of autonomous coaxial micro helicopters is given.

Chapter 2

Coaxial micro helicopter prototypes

In this section, the two coaxial micro helicopter prototypes developed for the muFly project are introduced. The focus is laid on the mechanical components and system integration, which pose significant challenges for the design of a fully autonomous helicopter with all necessary components on board. The ultimate goal is to fit the large variety of components with their unique constraints into one system with minimal mass.

In Section 2.1, the first working prototype of muFly is shown, which does not yet comprise the full set of sensors, but is able to perform partially autonomous flights. Designed as an initial test platform, the focus is more on modularity and on exchangeability of components than on integration. In Section 2.2, the second and final prototype of muFly is shown, which features the complete set of sensors and electronics for fully autonomous flight at significantly lower take off mass. There, the focus is certainly on integration. To quantify the mass reduction and effects of system integration, the two prototypes are quantitatively compared in Section 2.3.

2.1 Prototype muFly 1

The prototype muFly 1 has been developed as a general test platform. It allows for in-flight testing of system hardware and the attitude and altitude control algorithms [82] for the helicopter. The complete and assembled system is given in Figure 2.1, which shows that the helicopter features attitude and ground distance sensors, but no x-y-position sensor. The design goal of this prototype is to provide a robust platform with a low degree of integration and a high level of modularity to allow for easy testing in changing configurations. Due to the low level of integration, components can be easily exchanged for maintenance and repair, and new components can be added with only small modifications.

The main design constraints which need to be incorporated in the prototype are the following:

- Integration of the relatively heavy (11 g) and bulky (48 mm × 33 mm × 15 mm) standard IMU MTi OEM produced by Xsens (see Figure 2.2).

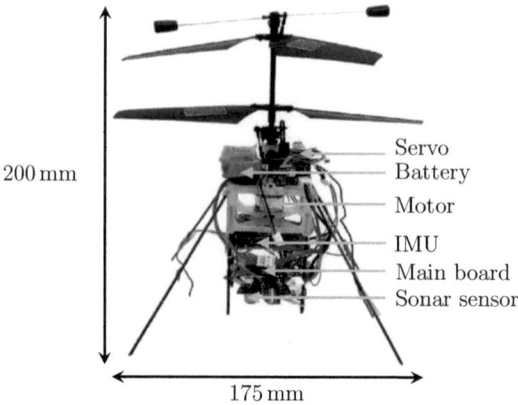

Figure 2.1: Assembled prototype muFly 1.

Figure 2.2: Xsens MTi OEM IMU for the prototype muFly 1 (Source: http://www.xsens.com).

- Free field of view for the ultrasonic rangefinder to measure the ground distance of the helicopter.

To meet these constraints, the helicopter is designed in two functional sections, which are a propulsion/drivetrain section and an electronics/sensor section. The propulsion/drivetrain section consists of a toy helicopter drivetrain and rotor system taken from the 5G6 coaxial helicopter by Walkera [9] including the servo motors. The rotors are driven by two Brushless Direct Current (BLDC) outrunner motors LRK 13-4-15Y [10], using 1:1.5 ratio spur gears.

All of these components are retained by a central structural part, which also serves as the connection to the electronics/sensor section and holds the landing gear. This central frame is produced in rapid prototyping and shown in Figure 2.3. The advantage of using Stereolithography rapid prototyping lies in the realizability of very complex three dimensional structures at a relatively low cost for small quantities. Moreover, the structure can be

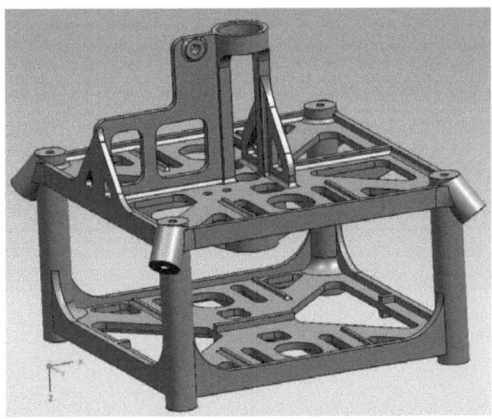

Figure 2.3: Central frame with receptacles for the sensor/electronics section and the landing gear as CAD model.

easily modified and manufactured within about three hours. On the downside, however, the properties of the rapid prototyping material are relatively poor, becoming more brittle with increasing age and with its E-modulus reduced by the temperature increase due to motor waste heat. These drawbacks have to be compensated by overdimensioning of the central frame. Therefore, the main structure is certainly not an optimum in mass, but a good compromise for a test platform that needs high adaptability to changing system components.

The electronics/sensor section consists of four carbon fiber rods in a square arrangement, which are pushed into receptacles at the bottom of the main frame and serve as a rack for all electronic components and sensors. Hence, the components can be pushed onto the rods and stacked. This allows for fixation of the IMU with dedicated adapters, as well as fixation of the main processing board, the motor controllers and the ultrasonic rangefinder at the bottom of the helicopter for a free field of view. This is also shown in Figure 2.1. A complete overview of the mass contributions of the individual components to the total mass of the helicopter of 95.84 g is given in Table 2.1. Inspection of Figure 2.1 and Table 2.1 clearly shows that the modularity and interchangeability of the helicopter are penalized by an increased total mass. The need for various cables and connectors, and the fact that the structural components of the helicopter have to be tailored to the electronic components and sensors, lead to large mass contributions of the structure and miscellaneous electronics, the latter mostly summarizing cables, connectors and additional components that cannot be integrated into the main electronic board. Also, with a mass of 11 g, the IMU contributes more then 10 % to the total helicopter mass. Since the maximum take-off mass lies at 100 g for the rotors and propulsion system used, the helicopter is fully loaded

Table 2.1: Mass distribution of the prototype muFly 1.

	Component	Comp. mass [g]	Quant. [-]	Total mass [g]
Structure	Main frame	7.83	1	7.83
	Carbon rod	0.44	4	1.76
	Motor holder	0.35	4	1.40
	Misc.	5.27	1	5.27
	Total mass structure			**16.26**
Sensors	IMU	11.00	1	11.00
	Ultrasonic rangefinder	3.75	1	3.75
	Total mass sensors			**14.75**
Electronics	Motor contr.	2.20	2	4.40
	Main board	8.30	1	8.30
	Misc.	6.70	1	6.70
	Total mass electronics			**19.40**
Propulsion	Coaxial shaft	4.04	1	4.04
	Blade	0.98	4	3.92
	Bearing	0.25	4	1.00
	Gear	0.19	4	0.76
	Stabilizer bar	2.75	1	2.75
	Total mass propulsion			**12.47**
Actuators	Motor	6.30	2	12.60
	Servo	3.83	2	7.66
	Total mass actuators			**20.26**
Battery		12.70	1	12.70
	Total mass helicopter			**95.84**

in this configuration and cannot carry further components, for instance the x-y-position sensor that is required for full position control. With 4 %, the remaining thrust margin is extremely small. Thus, in order to add further components to the helicopter, ways need to be found to reduce the total mass, since a significant increase of the total thrust of the system can not be expected.

2.2 Prototype muFly 2

With a largely defined rotor and propulsion system, the only way to reduce the mass of the helicopter and consequently allow for addition of further components, is a tighter integration of all the existing components. The key to this is dual use of as many components as possible, for instance by simultaneously using the necessary electronics as helicopter structure.

Since the design goal for the prototype muFly 2 is a level of integration that is as high as possible, and since for this prototype all sensors and electronics can be produced to the specific need in terms of geometry, the design constraints differ from those of the prototype muFly 1:

- Compulsory integration of the omnidirectional camera and the laser diodes as x-y-position sensor.
- Compliance with a minimal distance of 90 mm between the laser plane and the focal center of the camera to achieve optimal resolution of the distance measurement.
- High structural stiffness to minimize displacement of laser diodes and camera with respect to each other, hence ensuring the precision of the x-y-distance measurement.

To meet these constraints, the helicopter is designed in a similar fashion as the quadrotor platform in [80]: all system electronics are used dually as structural parts. Horizontally and vertically placed PCBs intermesh, such that a three dimensional 'puzzle' is established. This offers the main potential to save structural mass. Moreover, almost all electrical connections can be achieved by soldering the PCBs, making cables and connectors obsolete and offering another field of potential mass reduction, the electronics.

The complete and assembled prototype muFly 2 is shown in Figure 2.4. It can be seen that most of the helicopter is built up from plate-like structures that additionally serve as systems electronics. The only structural parts that do not serve a second purpose are a central plate to hold motors and servos, the landing gear, and bearing holders manufactured by rapid prototyping, which are necessary to hold the rotor shafts in position. These bearing holders serve as adaptors between the plate-like structures of the electronics and the cylindrical ball bearings. A complete overview of the disassembled prototype is shown in Figure 2.5, where it becomes more visible that the helicopter is mostly assembled from plate-like parts. In the figure, all components are placed in positions corresponding to their actual positions in the assembled version. It can be seen that cables are only needed

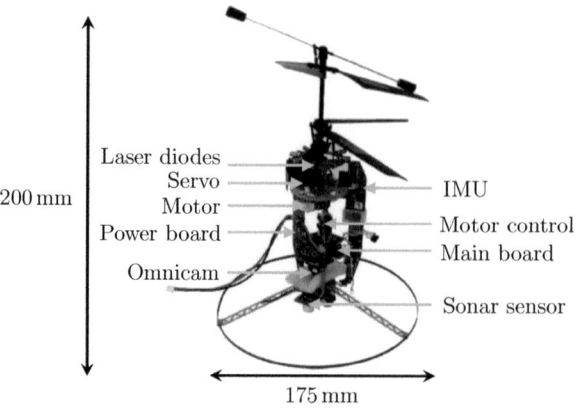

Figure 2.4: Assembled prototype muFly 2.

for the connections of the actuators and the battery, and that in general, the helicopter consists of a relatively small number of parts.

Principally, the helicopter is mounted around a horizontal central structural plate, which holds the motors and the servos. The three vertical PCBs are pushed onto the plate and hold the bearing holders for the drive train in place from three directions. Other horizontal components held by the three vertical PCBs are the main processing board and the omnidirectional camera. On top and bottom, the vertical PCBs are held together by the laser diode PCB and the PCB for the ultrasonic range finder, which are mounted as outer rings around the vertical PCBs. Hence, displacement of the vertical PCBs in the horizontal direction is prevented and a very stiff structure is achieved.

Of special interest, according to the design goals for this prototype, is the integration of a complete set of position sensors. These sensors are shown in Figure 2.6 with the laser diodes (left), omnidirectional camera for detection of the laser points (center) and ultrasonic range finder (right), the latter being the same as on the prototype muFly 1. The PCBs for the laser diodes and the omnidirectional camera are designed such that they can be integrated in the intermeshed three dimensional structure. A minimal distance of 90 mm must be obeyed between the laser plane and the focal center of the omnidirectional camera in order to ensure optimal resolution of the optical distance measurement. Since the lowermost position on the helicopter is already reserved for a free field of view of the ultrasonic rangefinder, the omnidirectional camera must be mounted higher than that. Therefore, with a height of 200 mm, the helicopter appears to be higher than necessary.

For this prototype of the muFly helicopter, a specialized BLDC outrunner motor has been developed by the project partner CEDRAT [4]. This motor is an improved version of the BLDC motor LRK 13-4-15Y that is used for the muFly 1 prototype. The motor

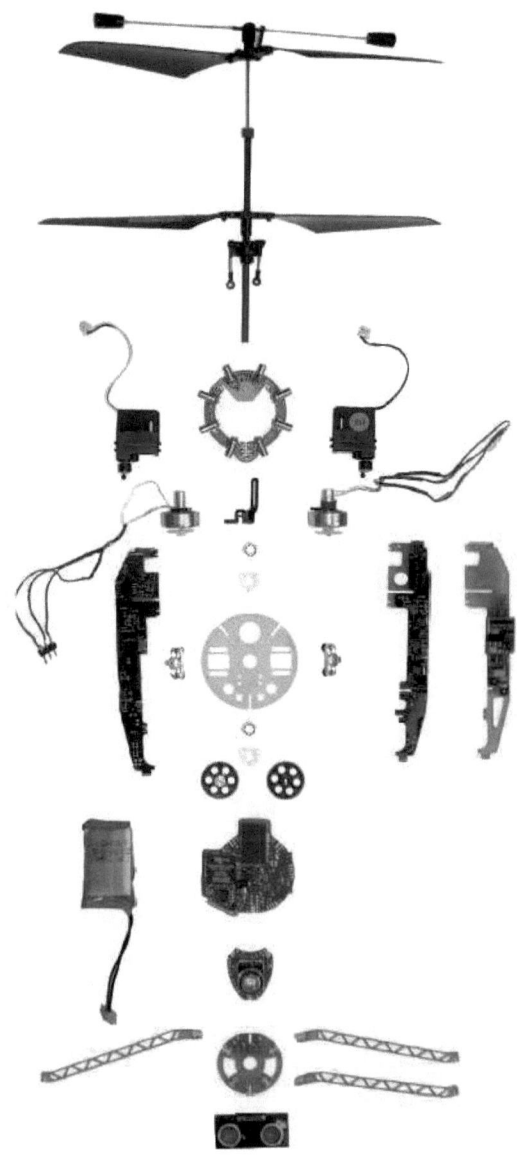

Figure 2.5: Disassembled prototype muFly 2.

Figure 2.6: Position sensors of the prototype muFly 2: combined laser diodes (left) and omnidirectional camera (center), ultrasonic range finder (right).

features silver wiring in an optimized arrangement for a higher motor efficiency at only slightly increased mass. This additional mass is mostly compensated by the fact that the motor includes its fixation, making additional motor holders as in the muFly 1 prototype obsolete.

A complete overview over the mass distribution of the prototype muFly 2 is given in Table 2.2. With a total mass of 80.31 g despite the added laser diodes and omnidirectional camera, the mass reduction of the second prototype becomes obvious. This increases the thrust margin of the helicopter to almost 20 %. Another important observation is that the mass contribution of the structure of the helicopter is very small, which is a result of the dual use of the electronic boards of the system as structural parts. The major mass fractions of the second prototype are battery and payload.

2.3 Prototype comparison

In this section, a quantitative comparison between the two muFly prototypes is made based on the mass data in Table 2.1 and Table 2.2.

The mass percentages of the main functional groups of the prototype muFly 1 are shown in Figure 2.7. The major contribution to the total mass of 95.84 g is made by the actuators, i.e. motors and servos, which consume almost one quarter of the total mass. Other significant contributions come from the electronics and the structure. In general, it would be desirable to have high mass percentages of the sensors, the battery and the electronics, because they can be considered to be useful payload either for autonomous flight or flight endurance. For the prototype muFly 1, however, these percentages are relatively low. This result reflects the modular design of muFly 1, where easy exchangeability of standard components is penalized by a high structural mass fraction and low battery, sensors and electronics mass fractions.

In Figure 2.8, the mass percentages of the main functional groups are shown for the prototype muFly 2. This prototype features a total mass of 80.31 g, approximately 15 g lighter

Table 2.2: Mass distribution of the prototype muFly 2.

	Component	Comp. mass [g]	Quant. [-]	Total mass [g]
Structure	Central plate	1.82	1	1.82
	Landing gear	1.34	1	1.34
	Misc.	0.63	1	0.63
	Total mass structure			**3.79**
Sensors	IMU	4.88	1	4.88
	Laser diodes	3.25	1	3.25
	Camera	3.57	1	3.57
	Ultrasonic rangefinder	3.75	1	3.75
	Total mass sensors			**15.45**
Electronics	Motor contr.	3.45	1	3.45
	Power board	3.92	1	3.92
	Main board	6.07	1	6.07
	US PCB	1.56	1	1.56
	Total mass electronics			**15.00**
Propulsion	Coaxial shaft	4.04	1	4.04
	Blade	0.98	4	3.92
	Bearing	0.20	2	0.40
	Gear	0.21	4	0.84
	Stabilizer bar	2.75	1	2.75
	Total mass propulsion			**11.95**
Actuators	Motor	6.88	2	13.76
	Servo	3.83	2	7.66
	Total mass actuators			**21.42**
Battery		12.70	1	12.70
	Total mass helicopter			**80.31**

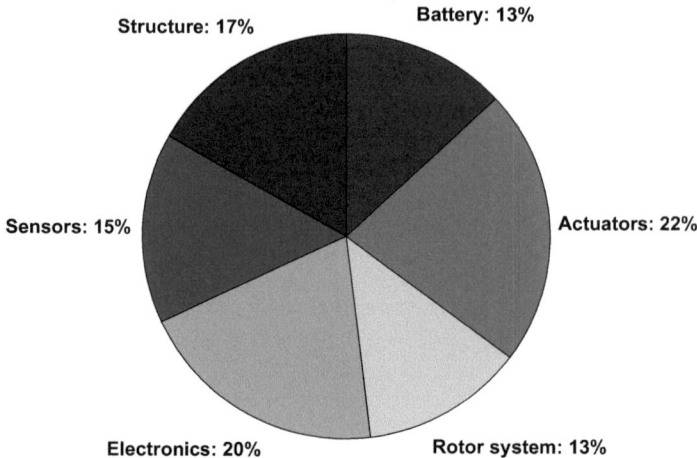

Figure 2.7: Mass distribution over main functional groups of the prototype muFly 1 at a total mass of 95.84 g.

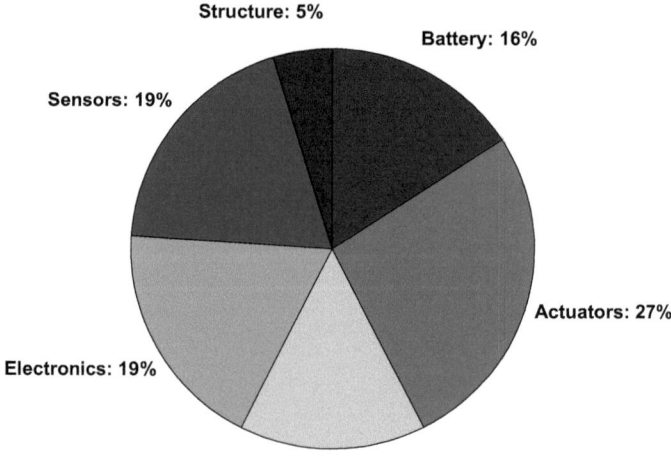

Figure 2.8: Mass distribution over main functional groups of the prototype muFly 2 at a total mass of 80.31 g.

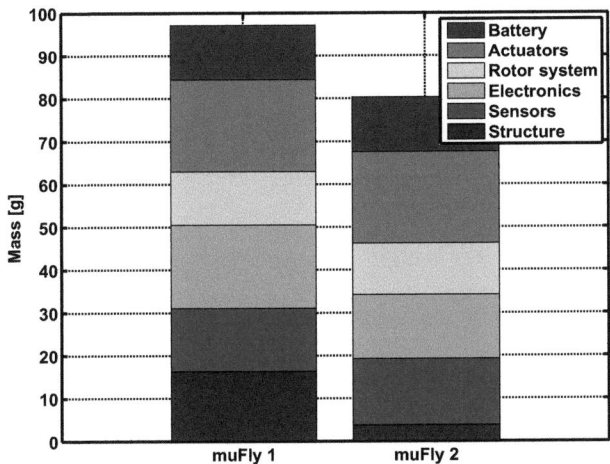

Figure 2.9: Mass comparison and distribution over main functional groups for the two prototypes of muFly.

than the first prototype. The most obvious change with respect to muFly 1 is the reduction of the structural mass, which could be suppressed to 5 % of the total mass. This is a result of the design goal of a strongly integrated helicopter that only uses specifically designed components, does not allow for easy exchange of components with standardized mechanical interfaces, and makes extensive use of dual purpose components, as the electronics, which are also used as structural parts. Another indicator for the success of dual use components is that the mass fraction of the electronics has not increased. At 19 %, it is even slightly lower than for the prototype muFly 1. The raised percentage for the actuators is a result of the slightly increased mass of the optimized motors , while the total mass of the helicopter has decreased. Since the battery and the rotor system are exactly the same as on prototype muFly 1, their percentages have also slightly increased. The increase in the sensor percentage is obviously explained by the additional mass that is introduced by the position sensor consisting of the omnidirectional camera and the laser diodes. Moreover, this increase can be considered desirable, since sensors are payload for autonomous flight. In total, the percentage of the useful payload of sensors, electronics and battery could be increased from 48 % on muFly 1 to 54 % on muFly 2.

Comparing the absolute mass values of the two prototypes, a similar result becomes visible. It is shown in Figure 2.9 for the complete helicopters, and in Figure 2.10 as a comparison of the functional groups. The significant mass reduction from 95.84 g to 80.31 g, which is a saving of approximately 17 %, is mainly achieved by reduction of the masses of structure

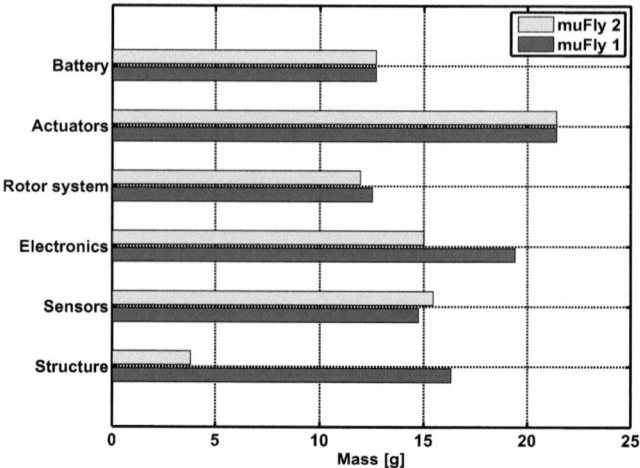

Figure 2.10: Mass comparison between the individual functional groups for the two prototypes of muFly.

and electronics.
Both prototypes are shown in successful test flight in Figure 2.11.

2.4 Summary

In this chapter, two prototype designs for the autonomous micro helicopter muFly are proposed and realized up to the level of successful test flight (see Figure 2.11). The prototypes are designed according to differing design goals and customized to the respective set of sensors they are intended to feature.

The first prototype is designed as a robust test helicopter for an early stage of the project,

Figure 2.11: Prototypes muFly 1 (left) and muFly 2 (right) in flight.

where not all sensors are available and where it has to be possible to use standard components with standard mechanical interfaces, like the MTi OEM IMU from Xsens. In general, the helicopter offers high modularity, allows for relatively fast exchange of components and is very robust. These advantages are countered by the disadvantage of a high structural and total mass of the helicopter, as the mass data analysis shows. Due to the low thrust margin of less than 4 %, mounting of an additional x-y-position sensor on the helicopter is not possible.

The second prototype is designed as a fully integrated system that features the complete set of sensors (IMU, x-, y- and z-position sensors) and the computational power necessary to allow for fully autonomous flight. By dual use of the electronic boards as structure, the structural mass is drastically reduced. Moreover, the structure is stiff enough to hold the combined laser-omnidirectional camera position sensor and guarantee its necessary mounting precision. The integrated design offers the additional advantage that cables and connectors become largely obsolete on the helicopter. Thus, despite the additional mass of the laser diodes and omnidirectional camera, the thrust margin of the helicopter is increased to almost 20 %. However, the integrated design does not allow for easy component exchange any more: all components must comply with the mechanical interfaces of the three dimensional structure made from plate-like elements.

In summary, also in the future only a very high degree of integration will lead to success for autonomous micro helicopters. On such small scales and with only low thrust margins, the comparatively heavy sensor package, which is oftentimes subjected to further design constraints like observation of a field of view, can only brought into the air if considered as an integral part of the helicopter, not an exchangeable add-on. This is a completely different paradigm as in the domain of UAVs, where oftentimes mission-specific pay load packages can be exchanged. For MAVs, the whole helicopter needs to be designed mission-specific.

Chapter 3
Conventions and basic concepts

For the remaining chapters of this thesis, it is necessary to introduce some conventions and basic concepts, which are valid throughout the course of the dynamic model derivations. They are summarized in this chapter.

Section 3.1 explains the basic functionality of the coaxial helicopter in terms of control of its six degrees of freedom. The two main coordinate systems and their rotational transformations are introduced in Section 3.2. This is followed by the definitions for the rotor azimuth and the degrees of freedom of a rotor blade in Section 3.3, description of rotor blade flapping in Section 3.4 and the resulting rotor moments in Section 3.5. Section 3.6 shows the basic equations of motion that are used for the helicopter simulation model.

3.1 Basic functionality of the coaxial helicopter

The basic functionality of the coaxial micro helicopter differs significantly from that of full scale helicopters in conventional main and tail rotor configuration, and also from that of full scale coaxial helicopters. In the following, the basic setup and control of the six degrees of freedom of the coaxial micro helicopter are explained.

The coaxial micro helicopter in the focus of this work consists of two rotors that are stacked on top of each other and rotate in opposite directions, as shown in Figure 3.1. Both rotors create a drag torque opposite to their direction of rotation, which must be balanced to zero to achieve a constant yaw angle of the helicopter. For a yawing motion of the helicopter, the speeds of the two rotors are differentially varied, to maintain constant thrust but achieve a drag torque imbalance that rotates the helicopter about its yaw axis. Also by speed variation of the two rotors, the altitude of the helicopter is changed: simultaneous reduction or increase of the two rotor speeds maintains a drag torque balance, but decreases or increases the total thrust of the helicopter, which leads to sinking or rising. In contrast to full scale conventional or coaxial helicopters, the coaxial micro helicopter does not feature a collective pitch swash plate [95], which changes the mean blade pitch angle and changes the thrust of the rotors without variation of their speeds.

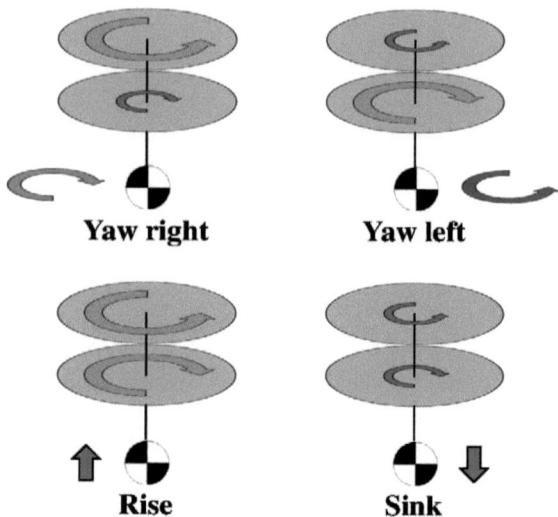

Figure 3.1: Yaw and vertical motion of the coaxial micro helicopter.

To achieve displacement of the helicopter in the body-fixed x- and y-directions, roll and pitch motions have to be initiated, which means that the horizontal translational degrees of freedom are coupled to the horizontal rotational degrees of freedom. Two ways of initiating roll and pitch motions are considered for the coaxial micro helicopter: on the one hand by using a cyclic pitch swash plate (see Figure 3.2 (top)), as it is also extensively used on full scale helicopters of all configurations, and on the other hand by a mechanism to displace the center of mass of the helicopter (see Figure 3.2 (bottom)). The cyclic pitch swash plate tilts the rotor disc by means of the aerodynamic forces, the fuselage then follows this tilt. The center of mass displacement mechanism tilts the fuselage, then the rotor disc follows that tilt. Both mechanisms, however, provide control over the horizontal attitude of the helicopter and subsequently the respective translational degrees of freedom.

3.2 Coordinate systems

Two coordinate systems are used to describe the position and attitude of the helicopter in space. These are the inertial coordinate frame $\{I\}$ and the body-fixed coordinate frame $\{B\}$. Both frames are right-handed, and while the X-, Y- and Z-axis of frame $\{I\}$ are oriented *North, East, Down*, the x-, y- and z-axis of frame $\{B\}$ are oriented *Forward, Right, Down* (see Figure 3.3). The inertial frame is earth-fixed, and the body-fixed frame is fixed in the center of mass of the helicopter. The rotational transformation

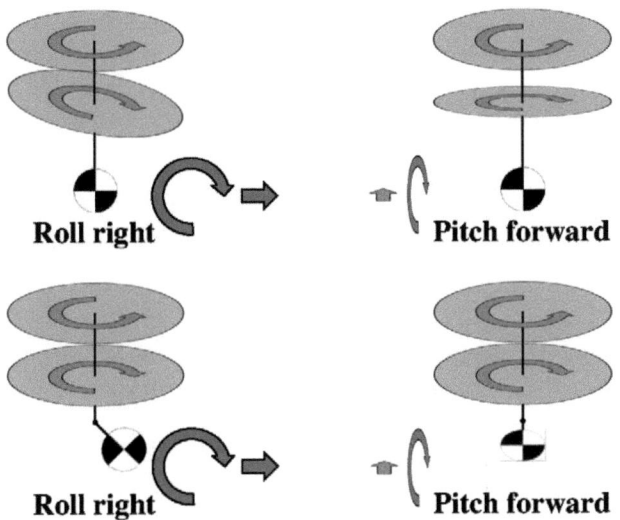

Figure 3.2: Roll and pitch motion of the coaxial micro helicopter for steering by a cyclic pitch swash plate on the lower rotor (top row) and by center of mass displacement (bottom row).

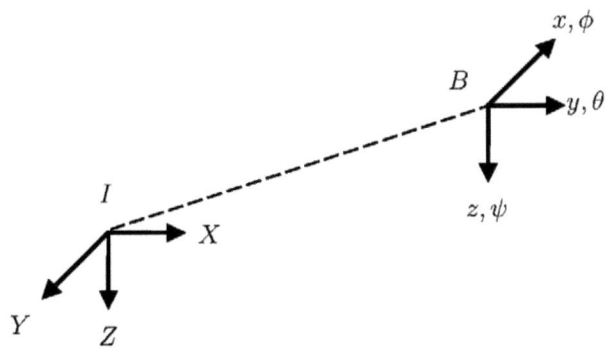

Figure 3.3: Inertial and body-fixed coordinate frames $\{I\}$ and $\{B\}$.

of a vector from the inertial frame $\{I\}$ to the body-fixed frame $\{B\}$ can be expressed as:

$$\begin{bmatrix} x \\ y \\ z \end{bmatrix} = \mathbf{A}_{BI} \begin{bmatrix} X \\ Y \\ Z \end{bmatrix}. \tag{3.1}$$

The rotational transformation matrix $\mathbf{A}_{BI}$ is defined by three consecutive rotations in roll (ϕ), pitch (θ) and yaw (ψ), following the Tait-Bryan convention [37]:

$$\mathbf{A}_{BI} = \begin{bmatrix} c\theta c\psi & c\theta s\psi & -s\theta \\ -c\phi s\psi + s\phi s\theta c\psi & c\phi c\psi + s\phi s\theta s\psi & s\phi c\theta \\ s\phi s\psi + c\phi s\theta c\psi & -s\phi c\psi + c\phi s\theta s\psi & c\phi c\theta \end{bmatrix}, \tag{3.2}$$

where s = sin and c = cos.
Since the previously defined transformation does not hold for angular quantities, these require a different transformation. Specifically, this transformation is defined for converting the time derivatives of the Tait-Bryan angles into body rotational velocities $[p, q, r]^T$:

$$\begin{bmatrix} p \\ q \\ r \end{bmatrix} = \mathbf{R}_{BI} \begin{bmatrix} \dot{\phi} \\ \dot{\theta} \\ \dot{\psi} \end{bmatrix}. \tag{3.3}$$

The matrix $\mathbf{R}_{BI}$ for transforming angular quantities from the inertial to the body-fixed frame is defined as:

$$\mathbf{R}_{BI} = \begin{bmatrix} 1 & 0 & -\sin\theta \\ 0 & \cos\phi & \sin\phi\cos\theta \\ 0 & -\sin\phi & \cos\phi\cos\theta \end{bmatrix}. \tag{3.4}$$

These transformations complete the coordinate system definitions.

3.3 Rotor azimuth and blade degrees of freedom

In agreement with the rotor directions of rotation on the helicopter prototypes built for this thesis, the directions of rotation of the upper and lower rotor are defined as shown in Figure 3.4: seen from above, the upper rotor rotates in the counterclockwise direction at a speed of Ω_{up}, and the lower rotor in the clockwise direction at a speed of Ω_{lo}. Their time integrals are the rotor azimuth angles $\Psi_{\{\text{up,lo}\}} = \Omega_{\{\text{up,lo}\}} t$. Since the both azimuth angles are defined as $\Psi_{\{\text{up,lo}\}} = 0°$ in the back of the helicopter, the azimuth $\Psi_{\text{up}} = 90°$ on the upper rotor corresponds to azimuth $\Psi_{\text{lo}} = 270°$ on the lower rotor and vice versa.
In general, helicopter rotor heads are designed articulated, which means that every blade has three rotational degrees of freedom at its root [95]. These rotational degrees of freedom are called feathering, flapping and lagging, according to Figure 3.5. While the feathering degree of freedom, which is called blade pitch throughout the course of this work, is needed

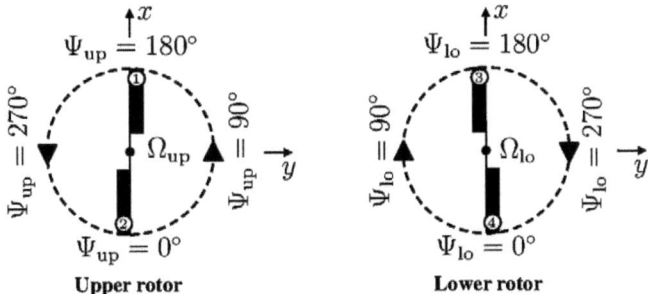

Figure 3.4: Schematic top view of the upper (left) and lower (right) rotor with directions of rotation and rotor azimuth angles $\Psi_{\{up,lo\}}$.

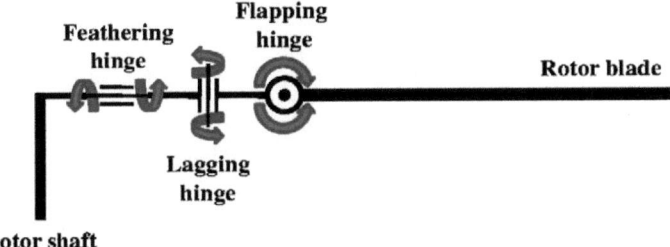

Figure 3.5: Degrees of freedom of an articulated rotor head.

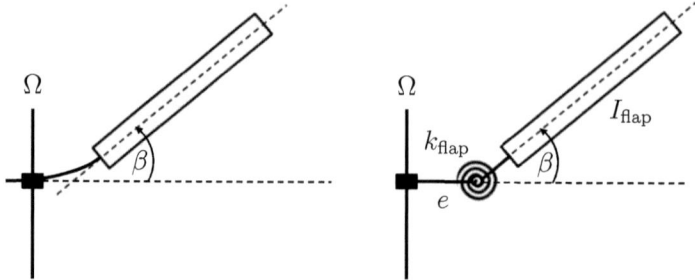

Figure 3.6: Schematics of the hingeless rotor with flexible blade root (left) and the analogous model with hinge offset e, flapping stiffness k_{flap}, blade flapping inertia I_{flap} and rotor angular speed Ω (right), adapted from [21].

for steering of the helicopter by changing the rotor blade pitch angles, flapping and feathering degrees of freedom are indispensable on a helicopter to compensate lift dissymmetry of the rotor in forward flight and its results. Further information on lagging is given in [77], while the following section explains the basic effect of blade flapping.

3.4 Rotor blade flapping

Rotor blade flapping describes the harmonic up-and-down motion of the rotor blades with respect to the rotor mast, either around a flapping hinge or by structural bending of the blade root. It is a result of active steering inputs, stabilizer bar inputs and roll, pitch and forward velocities of the helicopter. The flapping motion due to roll, pitch and forward velocities is defined as passively excited flapping, while flapping due to cyclic blade pitch variations is defined as active flapping. In principle, flapping is a forced oscillation of the rotor blades at the rotational frequency of the rotor shaft. Figure 3.6 shows the basic mechanic system considered, with a hingeless rotor blade (left), and its analogous model (right). In the analogous model, the blade is attached to the shaft rotating at the speed Ω through a hinge with hinge offset e and torsional stiffness k_{flap}. The blade itself has a flapping inertia of I_{flap}. The flapping angle β is a harmonic function of the form

$$\beta_{\text{up}} = -a_{1_{\text{up}}} \cos\left(\Omega_{\text{up}} t\right) - b_{1_{\text{up}}} \sin\left(\Omega_{\text{up}} t\right), \tag{3.5}$$

$$\beta_{\text{lo}} = -a_{1_{\text{lo}}} \cos\left(\Omega_{\text{lo}} t\right) + b_{1_{\text{lo}}} \sin\left(\Omega_{\text{lo}} t\right), \tag{3.6}$$

where the amplitudes $a_{1_{\{\text{up,lo}\}}}$ and $b_{1_{\{\text{up,lo}\}}}$ are the longitudinal and lateral projections of the flapping angle $\beta_{\{\text{up,lo}\}}$ on the body-fixed coordinate axes. Hence, the positive directions of these projections are defined as shown in Figure 3.7 for the upper and lower rotor. The schematic shows side view and front view of the helicopter to make the longitudinal and lateral flapping angles visible. For both rotors it is defined that the longitudinal flapping angles $a_{1_{\{up,lo\}}}$ are positive, when the blades flap up in front and down in the back of the helicopter, and that the lateral flapping angles $b_{1_{\{up,lo\}}}$ are positive, when the blades flap up to the left and down to the right of the helicopter. Positive flapping angles are shown for each of the respective two blades labeled $(1,2)$ and $(3,4)$ for both rotors.

Without taking aerodynamic or gyroscopic moments into account, the flapping equation of motion of a rotor blade according to Figure 3.7 can be derived. The moments acting around the flapping hinge are the inertial moment M^{in}, the centrifugal moment M^{cf} and the spring moment M^{flap}, which have to equilibrate:

$$M^{\text{in}} = \ddot{\beta} I_{\text{flap}}, \tag{3.7}$$

$$M^{\text{cf}} = \Omega^2 \beta (I_{\text{flap}} + e M_{\text{bld}}/g), \tag{3.8}$$

$$M^{\text{sprg}} = \beta k_{\text{flap}}, \tag{3.9}$$

$$M^{\text{in}} + M^{\text{cf}} + M^{\text{sprg}} = 0, \tag{3.10}$$

with the static blade moment

$$M_{\text{bld}} = \frac{3}{2} \frac{g I_{\text{flap}}}{R - e}. \tag{3.11}$$

The equilibrium yields a second order ordinary differential equation for the flapping angle β with the eigenfrequency

$$\omega_{\text{eig}} = \Omega \sqrt{1 + \frac{e M_{\text{bld}}}{g I_{\text{flap}}} + \frac{k_{\text{flap}}}{\Omega^2 I_{\text{flap}}}}. \tag{3.12}$$

Obviously, $e = 0$ and $k_{\text{flap}} = 0$ yields $\omega_{\text{eig}} = \Omega$, for all other meaningful (i.e. larger than zero) values of hinge offset e and spring stiffness k_{flap}, the eigenfrequency ω_{eig} becomes larger than the excitation frequency Ω. From Figure 3.8, which shows the phase delay for a damped oscillatory system, it becomes then clear that for a rotor without hinge offset and spring stiffness the phase delay must be $90°$, while it can only be smaller, if hinge offset or spring stiffness are introduced. In order to determine the exact phase delay, the damping of the system must be known. It is introduced by the aerodynamic forces.

In general, blade flapping on a helicopter stems from three sources:

- Roll and pitch velocities of the helicopter, leading to the passively excited flapping angles $a_{1_{\{up,lo\}}}^{\text{rot}}$ and $b_{1_{\{up,lo\}}}^{\text{rot}}$.
- Forward velocity of the helicopter, leading to the passively excited flapping angles $a_{1_{\{up,lo\}}}^{\text{fwd}}$ and $b_{1_{\{up,lo\}}}^{\text{fwd}}$.
- Steering inputs on the swash plate or stabilizer bar, leading to the actively excited flapping angles $a_{1_{\{up,lo\}}}^{\text{sp,sb}}$ and $b_{1_{\{up,lo\}}}^{\text{sp,sb}}$.

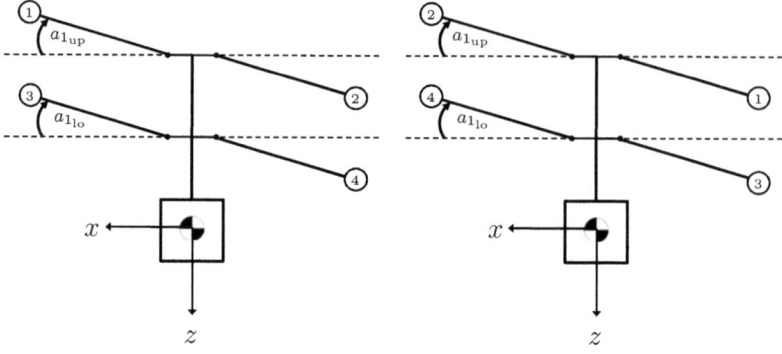

Longitudinal flapping (side view)

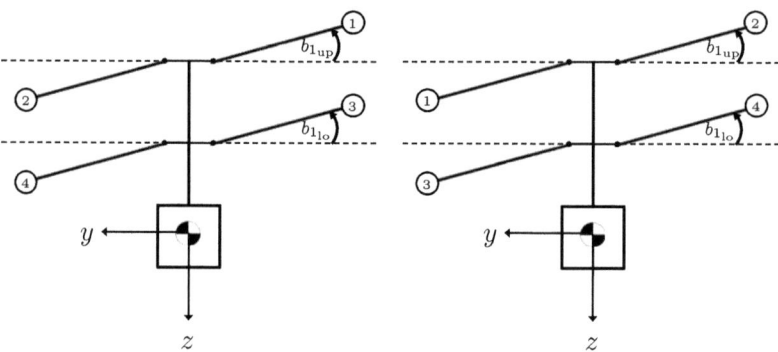

Lateral flapping (front view)

Figure 3.7: Schematic view of rotor blade flapping with longitudinal flapping angles $a_{1_{\{up,lo\}}}$ and lateral flapping angles $b_{1_{\{up,lo\}}}$.

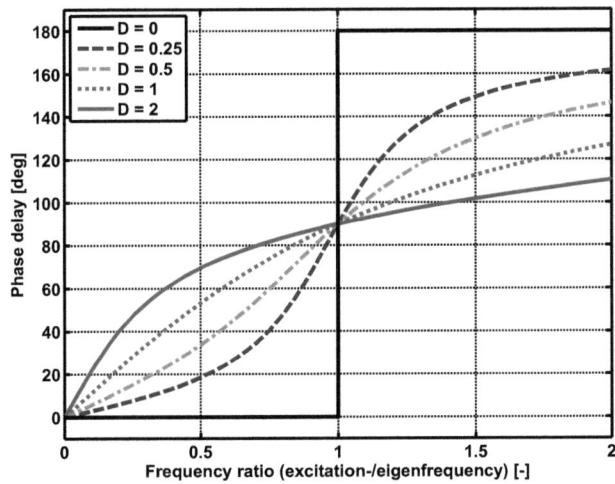

Figure 3.8: Phase delay of an oscillatory system as a function of frequency ratio and normalized damping D.

Conveniently, these flapping angles can be superimposed for their respective axis. Hence the longitudinal and lateral flapping angles $a_{1_{\{up,lo\}}}$ and $b_{1_{\{up,lo\}}}$ can be expressed as

$$a_{1_{\{up,lo\}}} = a_{1_{\{up,lo\}}}^{rot} + a_{1_{\{up,lo\}}}^{fwd} + a_{1_{\{up,lo\}}}^{sp,sb},$$
$$b_{1_{\{up,lo\}}} = b_{1_{\{up,lo\}}}^{rot} + b_{1_{\{up,lo\}}}^{fwd} + b_{1_{\{up,lo\}}}^{sp,sb}.$$
(3.13)

3.5 Steering moments

The hingeless rotor design is commonly used for all types of model scale helicopters and micro helicopters. On a hingeless rotor, the steering moments that result from rotor blade flapping are twofold: the first contribution stems from the tilt of the rotor tip path plane as a result of rotor blade flapping. Since the thrust vector is always oriented perpendicular to the tip path plane, it is tilted as well. Thus, its line of action does not intersect with the center of mass of the helicopter any more, resulting in a steering moment of

$$\vec{M}_{\{up,lo\}} = \vec{r}_{BR} \times \vec{T}_{\{up,lo\}},$$
(3.14)

where $\vec{r}_{BR}$ is the vector from the center of mass of the helicopter to the respective rotor. This contribution is shown schematically in Figure 3.9 (left).

The second contribution to the steering moment is a result of the spring stiffness and hinge offset of the analogous model of the hingeless rotor. With the flapping of a blade, a spring is loaded, and this spring moment is transmitted from the rotor head to the fuselage as a

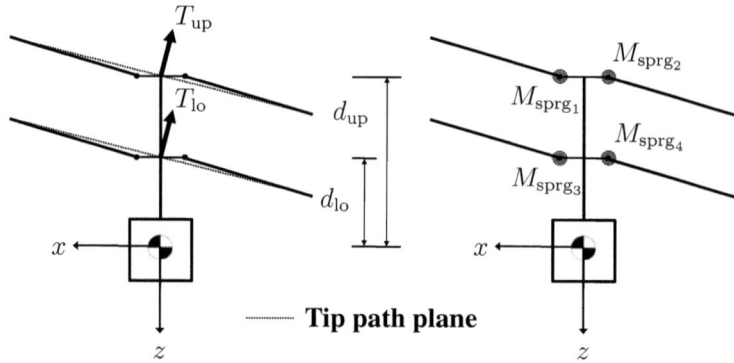

Figure 3.9: Two types of steering moments on the hingeless coaxial rotor.

second steering moment contribution. It can be formulated as

$$M_x = \left[\frac{1}{4}\frac{e}{R}bm_{\text{bld}}(\Omega R)^2 + bk_{\text{flap}}\right]b_1,$$
$$M_y = \left[\frac{1}{4}\frac{e}{R}bm_{\text{bld}}(\Omega R)^2 + bk_{\text{flap}}\right]a_1,$$
(3.15)

with the hinge offset e, rotor radius R, number of blades b, blade mass m_{bld}, rotor speed Ω, flapping stiffness k_{flap} and the longitudinal and lateral flapping angles a_1 and b_1. These moments correspond to the schematic display in Figure 3.9 (right). Note that like a flapping stiffness k_{flap}, a hinge offset e leads to a stiffer rotor. Since the steering moments on the coaxial micro helicopter come from two sources, it is interesting to see which of the sources has a larger influence, and how this influence scales. Thus, the respective steering moments are calculated for a set of baseline parameters (see Table 3.1) and their variation between ±100 %. The results of the parameter variations can be seen in Figure 3.10. For each variation, the moment from the rotor thrust, the moment from the rotor stiffness and their sum are shown. In the top left of Figure 3.10, the variation of the blade flapping angle is shown. Both moment contributions increase, the moment due to rotor stiffness, however, much faster. Due to the small distance between rotors and center of mass for micro helicopters, the moment due to thrust, which goes with the sine of the blade flapping angle, shows only a small variation. In the top right of Figure 3.10, the variation of the hinge offset is shown. Since the moment due to rotor thrust does only marginally depend on the hinge offset, it appears constant here, while the moment due to rotor stiffness increases with the hinge offset (growing rotor stiffness). In the bottom left of Figure 3.10, the flapping stiffness of the blades is varied. Again, the thrust moment is not affected by this parameter and remains at its baseline value, while the stiffness moment increases linearly. Finally, in the bottom right of Figure 3.10, the distance between the lower rotor and the center of mass of the helicopter is varied to influence the thrust moment alone.

Table 3.1: Baseline parameters for the steering moment variation.

	Parameter	Value	Unit	Description
Varied	a_1	5	[°]	Blade flapping angle
	e	0.01	[m]	Hinge offset
	k_{flap}	2.64×10^{-2}	[Nm/rad]	Blade flapping stiffness
	d_{lo}	0.04	[m]	Lower rotor distance from helicopter CM
Constant	b	2	[-]	Number of blades per rotor
	$d_{\text{up}} - d_{\text{lo}}$	0.4	[m]	Distance between the rotors
	m_{bld}	0.92×10^{-3}	[kg]	Blade mass
	R	0.0875	[m]	Rotor radius
	T	0.5	[N]	Single rotor thrust
	Ω	300	[rad/s]	Rotor angular speed

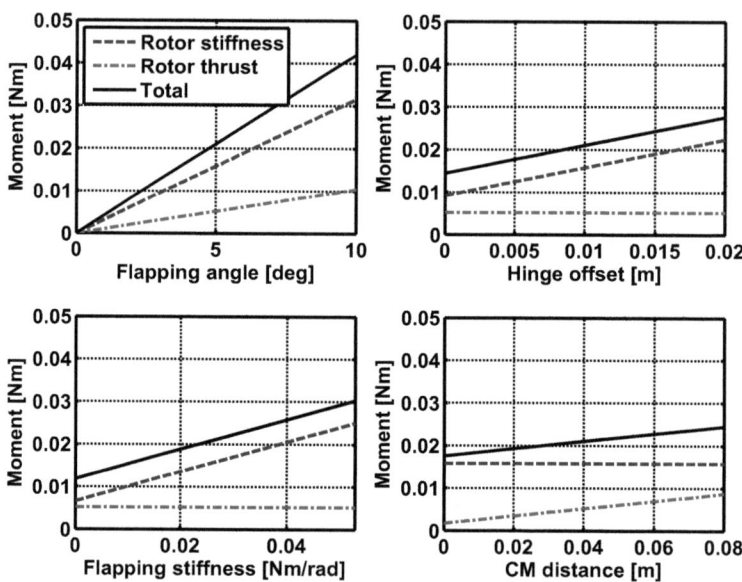

Figure 3.10: Variation of the steering moment from rotor stiffness and rotor thrust for a coaxial helicopter. Variation of the flapping angle (top left), hinge offset (top right), flapping stiffness (bottom left) and distance between lower rotor and helicopter center of mass (bottom right).

The thrust moments increases with increased distance, however, it never exceeds roughly 50 % of the stiffness moment with base line parameters.

These results lead to the conclusion that the stiffness moment is much more relevant for coaxial micro helicopters and that the influence of the thrust moment is comparably small. Hence, also the distance between lower rotor and center of mass of the helicopter as a design parameter is less relevant to the magnitude of the steering moments than the rotor related design parameters. However, large steering moments do not guarantee for roll and pitch stability of the helicopter. This question is further investigated in Chapter 6.

3.6 Helicopter equations of motion

The helicopter equations of motion are given here for completeness. Their derivation from Newtonian mechanics is shown in various references, for instance in [49].

$$\begin{aligned}
\sum F_x &= m\left(\ddot{x} - \dot{y}r + \dot{z}q\right) + mg\sin\theta, \\
\sum F_y &= m\left(\ddot{y} - \dot{z}p + \dot{x}r\right) - mg\cos\theta\sin\phi, \\
\sum F_z &= m\left(\ddot{z} - \dot{x}q + \dot{y}p\right) - mg\cos\theta\cos\phi, \\
\sum M_x &= I_{xx}\dot{p} - qr\left(I_{yy} - I_{zz}\right), \\
\sum M_y &= I_{yy}\dot{q} - pr\left(I_{zz} - I_{xx}\right), \\
\sum M_z &= I_{zz}\dot{r} - pq\left(I_{xx} - I_{yy}\right).
\end{aligned} \quad (3.16)$$

In the body-fixed coordinate frame, the external forces and moments are related to the body linear velocities $[u, v, w]^T$ and rotational velocities $[p, q, r]^T$ and their time derivatives. Note that Equation (3.16) already takes into account the gravitational force.

3.7 Summary

In this chapter, the conventions and basic concepts that are relevant for the theoretical part of this thesis are introduced. For the conventions, the relevant coordinate systems and their angular transformations are shown, and the definitions for longitudinal and lateral blade flapping angles are given. Furthermore, the equations of motion for a rigid body with six degrees of freedom, which are used as the equations of motion of the helicopter, are introduced.

In terms of basic concepts, the basic functionality of a coaxial micro helicopter is shown for its six degree of freedom control, the fundamental blade degrees of freedom are explained, and the concept of blade flapping with its resulting steering moments is introduced. These explanations include a sensitivity study of the steering moments with respect to relevant helicopter design parameters. The study clearly shows that the larger and more sensitive contribution to the overall steering moments stems from the rotor stiffness of the hingeless rotor, and not from the distance between the rotors and the center of mass of the helicopter.

Chapter 4

Stabilizer bar modeling and verification

The stabilizer bar is a passive stabilization device, which is commonly used on toy helicopters, MAVs, UAVs and even some full scale helicopters (Bell 47 and Bell UH-1). Despite its simple mechanical layout and its obvious functionality, strict design rules for the stabilizer bar cannot be found in the literature. This chapter is dedicated to modeling the stabilizer bar from first principles to achieve a realistic simulation model. The goal is to develop an experimentally validated model, which allows for the calculation of the following time and phase angle of the stabilizer bar as functions of the blade and stabilizer bar design parameters.

The chapter is organized as follows: in Section 4.1, the physical systems and possible design variations are introduced. Section 4.2 contains the complete derivation of a dynamic simulation model using the Euler-Lagrange formalism, and Section 4.3 shows the respective simulation results. In Section 4.4 and Section 4.5, the experimental setup for model validation and the signal processing procedures are introduced, while Section 4.6 shows the experimental results and how they compare to the simulation results.

4.1 Physical system

This section describes the basic mechanical layout of the stabilizer bar and distinguishes it from other variants of its design. The stabilizer bar is shown as a schematic sketch and in reality in Figure 4.1. It is basically a rod mounted on a revolute joint at the tip of the rotor mast, with a mass at each of its ends to create a flapping inertia. Via a spherical joint, a rigid push rod and another spherical joint the stabilizer bar is connected to the rotor blades, such that the flapping degree of freedom of the former is coupled to the feathering degree of freedom of the latter. It must be noted that the two blades of the rotor are coupled in their feathering motion, i.e. feathering up of one blade is coupled to feathering down of the other blade and vice versa. Hence, flapping of the stabilizer bar effects a cyclic

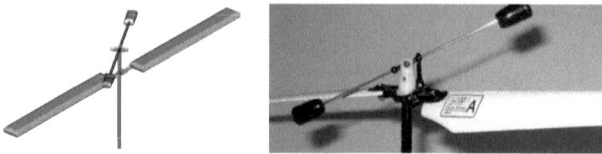

Figure 4.1: The stabilizer bar in schematics (left) and mounted on a model helicopter (right).

Figure 4.2: Definition of the phase angle α between blade feathering axis and stabilizer bar.

pitch input on the coupled rotor blades.

The design of the stabilizer bar is characterized by two relevant parameters, the stabilizer flapping inertia I_{sb} and the phase angle α. Assuming cylindrical masses and a cylindrical rod, and neglecting the push rod inertia, I_{sb} can be calculated as

$$I_{sb} = \frac{1}{12} m_{sbr} l_{sbr}^2 + 2 \left[\frac{1}{12} m_{sbm} \left(3 r_{sbm}^2 + l_{sbm}^2 \right) + m_{sbm} \left(l - \frac{l_{sbm}}{2} \right) \right], \quad (4.1)$$

where m_{sbr} and l_{sbr} are the mass and total length of the rod, and m_{sbm}, l_{sbm} and r_{sbm} are the mass, height and radius of each of the cylindrical end masses. Since the dimensions of the masses are small with respect to the length of the rod, the following approximation can be made with only a small error:

$$I_{sb} = \frac{1}{12} m_{sbr} l_{sbr}^2 + \frac{1}{2} m_{sbm} l_{sbr}^2. \quad (4.2)$$

The phase angle α is defined between the rotor blade feathering axis and the stabilizer bar, as shown in Figure 4.2. Correct tuning of the phase angle α is absolutely necessary. The maximal flapping angle of the blades has to occur at the same rotor azimuth as the maximal flapping angle of the stabilizer bar. If this is not the case, a disturbance on the roll *or* pitch axis will lead to blade flapping and therefore rotor moments on the roll *and* pitch axis, which is undesired cross coupling.

Originally, the stabilizer bar is an invention of Arthur M. Young, development engineer at Bell Helicopters. Based on his design, three basic variants of the system are known.

- The classical Bell stabilizer bar is a bar with cylindrical end masses (Figure 4.3 (left)), whose flapping degree of freedom is coupled to the feathering degree of freedom of the blades via push rods. Thus, cyclic actuation of the blades via the swash plate

Figure 4.3: The Bell stabilizer bar (left)(Source: http://www.zap16.com), the Hiller paddle auxiliary rotor (center)(Source: http://www.copters.com) and the Bell-Hiller system (right)(Source: http://www.rclineforum.de).

also actuates the stabilizer bar, resulting in a slower system response and higher flight stability. On an unsteered rotor, i.e. without a swash plate, stabilizer bar and rotor blades maintain their coupled flapping/feathering degree of freedom and act as a passive system.

- The Hiller paddle (see Figure 4.3 (center)) differs from the Bell stabilizer bar by using small airfoils instead of cylindrically shaped masses. Moreover, the paddles have a flapping and a feathering degree of freedom. The swash plate control inputs act on the feathering of the paddles alone, and the paddle flapping is coupled with the main blade feathering. Therefore, the paddle acts as an auxiliary rotor, exploiting aerodynamic forces to control the cyclic pitch on the main rotor blades. This can be of advantage for large rotors, which require large forces to adjust their blade pitch.

- Finally, the Bell-Hiller system (see Figure 4.3 (right)) is a variation of the Hiller paddle. Here, the swash plate input acts on the feathering degrees of freedom of both the paddle and the main rotor blades. The ratio of the two is governed by the mixing lever kinematics. The advantage of this system is that the dynamic response of the helicopter to steering inputs can be variably adjusted by changing the lever ratios. However, the system is mechanically much more complex than the original Bell stabilizer bar.

Neither the Hiller paddle nor the Bell-Hiller system will be analyzed in this work. A good reference on them is [35].

4.2 Euler-Lagrange modeling of the system

The Euler-Lagrange approach is an energy method to derive equations of motion for a mechanical system [49]. It is especially advantageous if a system consists of several bodies. In a classical Newton-Euler approach, the reaction forces and moments of the bodies with respect to each other would have to be calculated from a free-body-diagram. Instead, for the Euler-Lagrange approach the potential and kinetic energies for all bodies are formulated in terms of a set of generalized coordinates and all non-conservative forces acting on the system. Knowledge of the reaction forces and moments between the bodies is not required.

By solving the equation

$$\vec{a} = \frac{\mathrm{d}}{\mathrm{d}t}\left(\frac{\partial L}{\partial \vec{q}}\right)^{\mathrm{T}} - \frac{\partial L}{\partial \vec{q}}^{\mathrm{T}} - \vec{F}_{\mathrm{gen}} = 0, \qquad (4.3)$$

the equations of motion are obtained directly in terms of the generalized coordinates. In Equation (4.3), $L = T - V$ denotes the Lagrangian function, which is the difference between the total kinetic energy T and the total potential energy V of the system. The vector $\vec{q}$ contains the generalized coordinates, a minimal set of coordinates that fully describes the position of the system. All energy expressions are defined in terms of $\vec{q}$. In a generalized form, the vector $\vec{F}_{\mathrm{gen}}$ contains the non-conservative forces and moments acting on the system.

In the following section, the multi-body system at hand is introduced and the generalized coordinates are defined. Subsequently, the energy expressions are derived, and the external forces are introduced to complete the Lagrange formalism. Finally, the resultant equations of motion are linearized around a stationary solution. This linearization is then used for simulation. The approach for the problem at hand has been developed in [98].

4.2.1 System outline and coordinates

The system that is considered for the model is shown schematically in Figure 4.4. It consists of four rigid bodies, which are the fuselage, the two blades and the stabilizer bar. The rotor shaft and the link between stabilizer bar and blades are considered to be massless and are not included in the kinetic considerations, however, they are included as kinematic constraints. To describe the rotations of the four bodies with respect to each other, ten coordinate systems are defined:

{I}: The inertial coordinate frame, which remains earth-fixed.

{B}: The body coordinate frame, which is fixed on the fuselage. {B} is rotated with respect to {I} by the roll angle $\phi(t)$ and the pitch angle $\theta(t)$.

{T_1}: The shaft fixed coordinate frame, which follows the rotation of the shaft and points in the outward direction of the first blade. It is rotated with respect to {B} by the azimuth angle $\Psi(t) = \Omega t$.

{T_2}: The shaft fixed coordinate frame, which follows the rotation of the shaft and points in the outward direction of the second blade. It is rotated with respect to {B} by the azimuth angle $\Psi(t) = \Omega t + \pi$.

{T_S}: The shaft fixed coordinate frame, which follows the rotation of the shaft and points in the outward direction of the stabilizer bar. It is rotated with respect to {B} by the azimuth angle $\Psi(t) = \Omega t + \alpha$.

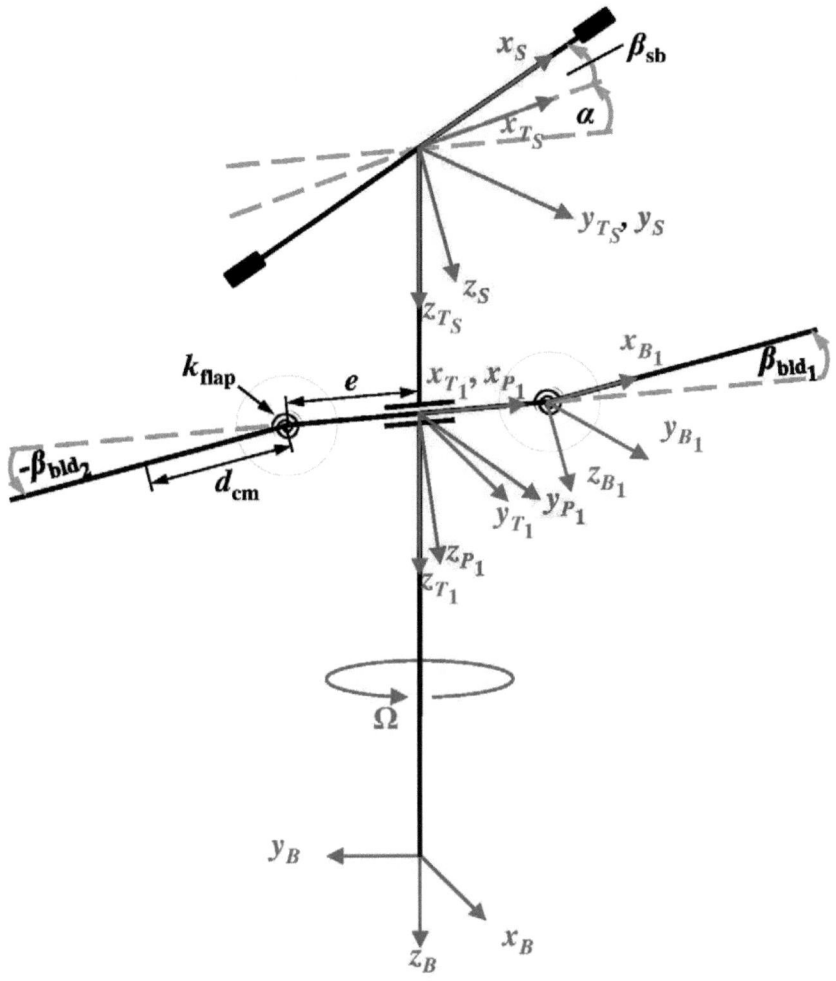

Figure 4.4: Schematic view of the multi-body system used for Euler-Lagrange modeling.

{**P₁**}: The pitched blade coordinate frame, which follows the pitching motion of the first blade. It is rotated with respect to $\{T_1\}$ by the blade pitching angle $\theta_\mathrm{p}(t)$.

{**P₂**}: The pitched blade coordinate frame, which follows the pitching motion of the second blade. It is rotated with respect to $\{T_2\}$ by the blade pitching angle $-\theta_\mathrm{p}(t)$.

{**B₁**}: The blade fixed coordinate frame of the first blade, which is rotated with respect to $\{P_1\}$ by the blade flapping angle $\beta_{\mathrm{bld}_1}(t)$.

{**B₂**}: The blade fixed coordinate frame of the second blade, which is rotated with respect to $\{P_2\}$ by the blade flapping angle $\beta_{\mathrm{bld}_2}(t)$.

{**S**}: The stabilizer bar fixed coordinate system $\{S\}$, which follows the flapping motion of the stabilizer bar. It is rotated with respect to $\{T_S\}$ by the stabilizer bar flapping angle $\beta_\mathrm{sb}(t)$.

Accordingly, the rotational transformation matrices between the inertial system and the four bodies can be defined:

- Inertial frame $\{I\}$ to fuselage fixed frame $\{B\}$:

$$\mathbf{A}_{BI} = \mathbf{R}_y(\theta)\mathbf{R}_x(\phi). \tag{4.4}$$

- Inertial frame $\{I\}$ to blade fixed frame $\{B_1\}$ of the first blade:

$$\mathbf{A}_{B_1 I} = \mathbf{R}_y(\beta_{\mathrm{bld}_1})\mathbf{R}_x(\theta_\mathrm{p})\mathbf{R}_z(\Psi)\mathbf{R}_y(\theta)\mathbf{R}_x(\phi). \tag{4.5}$$

- Inertial frame $\{I\}$ to blade fixed frame $\{B_2\}$ of the second blade:

$$\mathbf{A}_{B_2 I} = \mathbf{R}_y(\beta_{\mathrm{bld}_2})\mathbf{R}_x(-\theta_\mathrm{p})\mathbf{R}_z(\Psi + \pi)\mathbf{R}_y(\theta)\mathbf{R}_x(\phi). \tag{4.6}$$

- Inertial frame $\{I\}$ to stabilizer bar fixed frame $\{S\}$:

$$\mathbf{A}_{SI} = \mathbf{R}_y(\beta_\mathrm{sb})\mathbf{R}_z(\Psi + \alpha)\mathbf{R}_y(\theta)\mathbf{R}_x(\phi). \tag{4.7}$$

The corresponding matrices $\mathbf{R}_{\{x,y,z\}}$ for these transformations are given in Appendix B.
To define the translations of the four bodies with respect to each other, the following dimensions are introduced:

- The position h of the hub along the z_B-axis, which defines the distance between rotor hub and fuselage.
- The hinge offset e along the x_{P_1}- and x_{P_2}-axis, which is the perpendicular distance from the rotor hub to the respective flapping hinges of the blades.
- The position of the mass center of the blade d_cm along the x_{B_1}- and x_{B_2}-axis, which is the distance between flapping hinge and blade center of mass along the blade fixed x-axis.

Moreover, a rotational spring with stiffness k_{flap} is mounted in the blade flapping hinge to simulate structural stiffness of a hingeless rotor blade. As a simplification, it is assumed that the stabilizer bar and the rotor blades lie in the same plane. While this does not correspond to the physical system, introducing the offset between the two would not contribute to a more exact model.

Furthermore, it needs to be noted that the pitching angles of the two blades and the flapping angle of the stabilizer bar are kinematically coupled: blade pitching of one blade leads to an equal and opposite pitch of the other, such that

$$\theta_{\text{p}_1}(t) = -\theta_{\text{p}_2}(t). \tag{4.8}$$

Moreover, both angles depend linearly on the stabilizer bar flapping angle $\beta_{\text{sb}}(t)$. Thus, by using a gearing ratio κ it can be stated:

$$\theta_{\text{p}_1}(t) = \kappa \beta_{\text{sb}}(t). \tag{4.9}$$

With these definitions, the set of generalized coordinates can be introduced. It is a minimal set of coordinates to fully describe the position of the system, and corresponds to the degrees of freedom that are not restricted by bindings or imprinted motion. Since the pitching angles of the two blades and the flapping angle of the stabilizer bar are kinematically coupled according to Equations (4.8)–(4.9), they can be treated in one coordinate, the stabilizer bar angle $\beta_{\text{sb}}(t)$. Considering the rotor speed and thus the azimuth angle as imprinted, the following set of generalized coordinates $\vec{q}$ remains:

$$\vec{q} = \begin{bmatrix} \phi(t) \\ \theta(t) \\ \beta_{\text{sb}}(t) \\ \beta_{\text{bld}_1}(t) \\ \beta_{\text{bld}_2}(t) \end{bmatrix}. \tag{4.10}$$

Using the coordinates in this vector, the energy expressions are derived in the following section.

4.2.2 Kinetic and potential energy

For the evaluation of the Lagrange equation (4.3), expressions for the kinetic energy T and potential energy V of the system are needed.

Starting with the kinetic energy, its contributions are the rotational and the translational kinetic energy of each body. For the present system it is convenient to express the rotational energies of all bodies in their respective body-fixed coordinate system. By doing that, the inertia tensors expressed in the body-fixed systems can be used directly, which simplifies the calculations. As a further simplification it is assumed that the inertia tensors of all four bodies expressed in their respective body-fixed frame are diagonal:

$$\mathbf{I} = diag\left([I_{xx}, I_{yy}, I_{zz}]^T\right). \tag{4.11}$$

To find the rotational velocities of all bodies in their body-fixed frames, time derivatives of the transposed transformation matrices are needed. By left-multiplication with the transformation matrices, skew symmetric expressions $_B\tilde{\omega}_{BI}$, $_S\tilde{\omega}_{SI}$, $_{B_1}\tilde{\omega}_{B_1I}$ and $_{B_2}\tilde{\omega}_{B_2I}$ for the rotational velocities are found:

$$_B\tilde{\omega}_{BI} = \mathbf{A}_{BI} \cdot \frac{\mathrm{d}}{\mathrm{d}t}\left(\mathbf{A}_{BI}\right)^T, \tag{4.12}$$

$$_S\tilde{\omega}_{SI} = \mathbf{A}_{SI} \cdot \frac{\mathrm{d}}{\mathrm{d}t}\left(\mathbf{A}_{SI}\right)^T, \tag{4.13}$$

$$_{B_1}\tilde{\omega}_{B_1I} = \mathbf{A}_{B_1I} \cdot \frac{\mathrm{d}}{\mathrm{d}t}\left(\mathbf{A}_{B_1I}\right)^T, \tag{4.14}$$

$$_{B_2}\tilde{\omega}_{B_2I} = \mathbf{A}_{B_2I} \cdot \frac{\mathrm{d}}{\mathrm{d}t}\left(\mathbf{A}_{B_2I}\right)^T. \tag{4.15}$$

In general, the angular velocity vector $\vec{\omega}$ can be found from the skew symmetric velocity matrix $\tilde{\omega}$ via the relation

$$\vec{\omega} = \begin{bmatrix} \tilde{\omega}_{32} \\ \tilde{\omega}_{13} \\ \tilde{\omega}_{21} \end{bmatrix} = \begin{bmatrix} -\tilde{\omega}_{23} \\ -\tilde{\omega}_{31} \\ -\tilde{\omega}_{12} \end{bmatrix}. \tag{4.16}$$

Thus, using the skew symmetric matrices (4.12)–(4.15) and inertia tensors of the form (4.11), the rotational energies of the four bodies can be formulated as

$$T^{\mathrm{rot}}_{\mathrm{fus}} = \frac{1}{2} \cdot \left(_B\vec{\omega}_{BI}\right)^T \cdot \mathbf{I}_{\mathrm{fus}} \cdot {_B}\vec{\omega}_{BI}, \tag{4.17}$$

$$T^{\mathrm{rot}}_{\mathrm{sb}} = \frac{1}{2} \cdot \left(_S\vec{\omega}_{SI}\right)^T \cdot \mathbf{I}_{\mathrm{sb}} \cdot {_S}\vec{\omega}_{SI}, \tag{4.18}$$

$$T^{\mathrm{rot}}_{\mathrm{bld}_1} = \frac{1}{2} \cdot \left(_{B_1}\vec{\omega}_{B_1I}\right)^T \cdot \mathbf{I}_{\mathrm{bld}} \cdot {_{B_1}}\vec{\omega}_{B_1I}, \tag{4.19}$$

$$T^{\mathrm{rot}}_{\mathrm{bld}_2} = \frac{1}{2} \cdot \left(_{B_2}\vec{\omega}_{B_2I}\right)^T \cdot \mathbf{I}_{\mathrm{bld}} \cdot {_{B_2}}\vec{\omega}_{B_2I}. \tag{4.20}$$

The total rotational energy of the system is the sum of these four contributions:

$$T^{\mathrm{rot}} = T^{\mathrm{rot}}_{\mathrm{fus}} + T^{\mathrm{rot}}_{\mathrm{sb}} + T^{\mathrm{rot}}_{\mathrm{bld}_1} + T^{\mathrm{rot}}_{\mathrm{bld}_2}. \tag{4.21}$$

For the calculation of the translational kinetic energy of the system, first of all, the vectors from the inertial frame $\{I\}$ to the respective body mass centers expressed in the body-fixed frame need to be found. While this is obvious for the fuselage and the stabilizer bar, coordinate transformations are required for the two blades:

$$_B\vec{r}_{\text{cm},I} = \begin{bmatrix} 0 \\ 0 \\ 0 \end{bmatrix}, \tag{4.22}$$

$$_S\vec{r}_{\text{cm},I} = \begin{bmatrix} 0 \\ 0 \\ -h \end{bmatrix}, \tag{4.23}$$

$$_{B_1}\vec{r}_{\text{cm},I} = \begin{bmatrix} d_{\text{cm}} \\ 0 \\ 0 \end{bmatrix} + \mathbf{R}_y(\beta_{\text{bld}_1}) \cdot \left(\begin{bmatrix} e \\ 0 \\ 0 \end{bmatrix} + \mathbf{R}_z(\Psi) \cdot \begin{bmatrix} 0 \\ 0 \\ -h \end{bmatrix} \right), \tag{4.24}$$

$$_{B_2}\vec{r}_{\text{cm},I} = \begin{bmatrix} d_{\text{cm}} \\ 0 \\ 0 \end{bmatrix} + \mathbf{R}_y(\beta_{\text{bld}_2}) \cdot \left(\begin{bmatrix} e \\ 0 \\ 0 \end{bmatrix} + \mathbf{R}_z(\Psi + \pi) \cdot \begin{bmatrix} 0 \\ 0 \\ -h \end{bmatrix} \right). \tag{4.25}$$

The time derivatives of these four vectors yield the translational velocities of the bodies in their respective frames. For this, however, Euler's rule needs to be observed, which again introduces the rotational velocities (4.14)–(4.15) of the frames with respect to each other.

$$_B\vec{v}_{\text{cm},I} = 0, \tag{4.26}$$

$$_S\vec{v}_{\text{cm},I} = 0, \tag{4.27}$$

$$_{B_1}\vec{v}_{\text{cm},I} = \frac{\partial}{\partial t} \left(_{B_1}\vec{r}_{\text{cm},I}\right) +_{B_1} \tilde{\omega}_{B_1 I} \cdot _{B_1}\vec{r}_{\text{cm},I}, \tag{4.28}$$

$$_{B_2}\vec{v}_{\text{cm},I} = \frac{\partial}{\partial t} \left(_{B_2}\vec{r}_{\text{cm},I}\right) +_{B_2} \tilde{\omega}_{B_2 I} \cdot _{B_2}\vec{r}_{\text{cm},I}. \tag{4.29}$$

Thus, using the blade mass m_{bld}, the translational kinetic energies are formulated. Neither fuselage nor stabilizer bar contribute to this energy.

$$T^{\text{trans}}_{\text{fus}} = 0, \tag{4.30}$$

$$T^{\text{trans}}_{\text{sb}} = 0, \tag{4.31}$$

$$T^{\text{trans}}_{\text{bld}_1} = \frac{1}{2} \cdot \left(_{B_1}\vec{v}_{\text{cm},I}\right)^T \cdot m_{\text{bld}} \cdot _{B_1}\vec{v}_{\text{cm},I}, \tag{4.32}$$

$$T^{\text{trans}}_{\text{bld}_2} = \frac{1}{2} \cdot \left(_{B_2}\vec{v}_{\text{cm},I}\right)^T \cdot m_{\text{bld}} \cdot _{B_2}\vec{v}_{\text{cm},I}. \tag{4.33}$$

The sum of all non-zero translational energies is the total translational energy of the system:

$$T^{\text{trans}} = T^{\text{trans}}_{\text{bld}_1} + T^{\text{trans}}_{\text{bld}_2}. \tag{4.34}$$

Also, the sum of total rotational and total translational energy is the total kinetic energy T of the system:

$$T = T^{\text{rot}} + T^{\text{trans}}. \tag{4.35}$$

The formulation of the potential energy contributions is significantly simpler than for the kinetic energy. For simplification, gravity is not considered in the model. This assumption is justified since the restoring effect of the stabilizer bar does not depend on the gravitational field. The stabilizer bar reacts to changes in the roll and pitch rates, but does not have a 'sense' for the absolute zero position of the roll and pitch angles. Without the influence of gravity, the only potential energy in the system is stored in the spring stiffness of the two blades. Thus,

$$V_{\text{bld}_1} = \frac{1}{2} k_{\text{flap}} \left(\beta_{\text{bld}_1} \right)^2, \tag{4.36}$$

$$V_{\text{bld}_2} = \frac{1}{2} k_{\text{flap}} \left(\beta_{\text{bld}_2} \right)^2, \tag{4.37}$$

are the potential energies of each of the blades, assuming that they have an equal spring stiffness k_{flap}. The sum of these energies is the total potential energy V of the system:

$$V = V_{\text{bld}_1} + V_{\text{bld}_2}. \tag{4.38}$$

With that, complete formulations for the energy term L in the Lagrange equation (4.3) are defined.

4.2.3 External forces and moments

The Euler-Lagrange method allows to introduce external forces and moments separately from the energy expressions. While the basic intention is to account for non-conservative forces and moments in the system, in general any force or moment can be introduced at this point. The only restriction is that the forces and moments have to be expressed in terms of the generalized coordinates $\vec{q}$, which is done by multiplying them with Jacobian matrices.

For the introduction of an arbitrary force vector $_G\vec{F}_P$ expressed in the coordinate system $\{G\}$, the Jacobian with respect to the generalized coordinates $\vec{q}$ at any point P on the body is defined as:

$$_G\mathbf{J}_P^f\left(\dot{\vec{q}}\right) = \frac{\partial}{\partial \dot{\vec{q}}} (_G\vec{v}_P). \tag{4.39}$$

Similarly, for the introduction of an arbitrary moment vector $_G\vec{T}_P$ expressed in the coordinate system $\{G\}$, the Jacobian with respect to the generalized coordinates $\vec{q}$ at any point P on the body is defined as:

$$_G\mathbf{J}_P^t\left(\dot{\vec{q}}\right) = \frac{\partial}{\partial \dot{\vec{q}}} (_G\vec{\omega}_{GI}). \tag{4.40}$$

Therefore, the term $\vec{F}_{\text{gen}}$ in the Lagrange equation (4.3) can be replaced by

$$\vec{F}_{\text{gen}} = \left[_G\mathbf{J}_P^f\left(\dot{\vec{q}}\right)\right]^T \cdot{}_G \vec{F}_P + \left[_G\mathbf{J}_P^t\left(\dot{\vec{q}}\right)\right]^T \cdot{}_G \vec{T}_P \tag{4.41}$$

to introduce arbitrary forces and moments.

The only forces, which need to be introduced, are the aerodynamic forces acting on each of the blades. They are formulated as proportional with respect to the square of the rotational speed Ω, where the lift and drag parameters k^{lift} and k^{drag} are functions of the blade pitch angle θ_{p}. Thus,

$$\vec{F}^{\text{aero}}_{\text{bld}_{\{1,2\}}} = \begin{bmatrix} 0 \\ k^{\text{lift}}\Omega^2 \\ k^{\text{drag}}\Omega^2 \end{bmatrix} = \begin{bmatrix} 0 \\ k^{\text{lift}*}\Omega^2\theta_{\text{p}} \\ k^{\text{drag}*}\Omega^2\theta_{\text{p}} \end{bmatrix}. \tag{4.42}$$

Other external forces and moments do not need to be included in the model.

With the introduction of external forces and moments, the formulation of Equation (4.3) is complete and the equations of motion can be computed.

4.2.4 Stationary solution and linearization

The solution of the Lagrange Equation (4.3) for the given system leads to a set of five nonlinear differential equations. These equations are rather long and too complicated to be displayed in a meaningful manner. Also, computations with these equations consume an inappropriate amount of computational time and work space. In order to obtain a set of equations that is compact enough to be actually worked with, the five equations are linearized around a stationary solution in the following.

To find a stationary solution, all time derivatives of the generalized coordinates $\vec{q}$ must be equal to zero:

$$\ddot{\vec{q}} = \dot{\vec{q}} = \vec{0}. \tag{4.43}$$

By setting roll angle ϕ, pitch angle θ and stabilizer bar flapping angle β_{sb} to zero and solving the remaining terms of the nonlinear equations of motion for the blade flapping angles β_{bld_1} and β_{bld_2}, the stationary solution for the flapping angles is found to be

$$\beta_{\text{bld}_1}\big|_0 = \beta_{\text{bld}_2}\big|_0 = \frac{-k^{\text{lift}*}\Omega^2\theta_{\text{p}}d_{\text{cm}}}{k_{\text{flap}} + \Omega^2\left[I_{\text{bld},zz} - I_{\text{bld},xx} + m_{\text{bld}}d_{\text{cm}}(d_{\text{cm}} + e)\right]}. \tag{4.44}$$

With the appropriate parametrization (see Table 4.1), this corresponds to an upward flapping angle of the blades of 0.52°, which is the coning angle of the rotor and very small. Thus, the stationary solution is approximated as

$$\vec{q}\,|_0 = \begin{bmatrix} 0 \\ 0 \\ 0 \\ 0 \\ 0 \end{bmatrix}. \tag{4.45}$$

With the set of nonlinear equations of motion and the stationary solution $\vec{q}\,|_0$, a Taylor expansion around the stationary solution can be found by rewriting $\vec{q}$ and its time derivatives

Table 4.1: Parametrization of the stabilizer bar model.

Parameter	Value	Unit	Description
d_{cm}	3.36×10^{-2}	[m]	Blade mass center from hinge
e	1.50×10^{-2}	[m]	Flapping hinge offset
h	9.20×10^{-2}	[m]	Shaft length
k_{flap}	2.64×10^{-2}	[Nm/rad]	Blade flapping stiffness
k^{lift*}	3.80×10^{-8}	[Ns^2/rad^2/°]	Lift coefficient
k^{drag*}	1.81×10^{-12}	[Ns^2/rad^2/°]	Drag coefficient
m_{bld}	9.60×10^{-4}	[kg]	Blade mass
$I_{bld,xx}$	2.68×10^{-8}	[kgm^2]	Blade pitching inertia
$I_{bld,yy}$	4.67×10^{-7}	[kgm^2]	Blade lagging inertia
$I_{bld,zz}$	4.41×10^{-7}	[kgm^2]	Blade flapping inertia
κ	1	[-]	Gearing ratio

as stationary solutions and small deviations:

$$\begin{aligned} \vec{q} &= \vec{q}\,|_s + \delta \vec{q}, \\ \dot{\vec{q}} &= \dot{\vec{q}}\,|_s + \delta \dot{\vec{q}}, \\ \ddot{\vec{q}} &= \ddot{\vec{q}}\,|_s + \delta \ddot{\vec{q}}. \end{aligned} \qquad (4.46)$$

Then, the expansion becomes

$$\vec{a} = \vec{a}\,|_s + \left[\frac{\partial \vec{a}}{\partial \ddot{\vec{q}}}\right]_s \cdot \delta \ddot{\vec{q}} + \left[\frac{\partial \vec{a}}{\partial \dot{\vec{q}}}\right]_s \cdot \delta \dot{\vec{q}} + \left[\frac{\partial \vec{a}}{\partial \vec{q}}\right]_s \cdot \delta \vec{q} - \vec{F}_{\text{gen}} + \mathcal{O}(2). \qquad (4.47)$$

Since $\vec{a}\,|_s = 0$, and since the higher order terms $\mathcal{O}(2)$ are neglected, Equation (4.47) can be rewritten in the standard form

$$\mathbf{M}(t) \cdot \ddot{\vec{q}} + \mathbf{D}(t) \cdot \dot{\vec{q}} + \mathbf{C}(t) \cdot \vec{q} - \vec{F}_{\text{gen}} = \vec{0} \qquad (4.48)$$

with the symmetric mass matrix $\mathbf{M}(t)$

$$\mathbf{M}(t) = \begin{bmatrix} M_{11} & M_{12} & M_{13} & M_{14} & M_{15} \\ M_{12} & M_{22} & M_{23} & M_{24} & M_{25} \\ M_{13} & M_{23} & M_{33} & 0 & 0 \\ M_{14} & M_{24} & 0 & M_{44} & 0 \\ M_{15} & M_{25} & 0 & 0 & M_{55} \end{bmatrix}, \qquad (4.49)$$

the damping matrix $\mathbf{D}(t)$

$$\mathbf{D}(t) = \Omega \begin{bmatrix} D_{11} & D_{12} & D_{13} & D_{14} & D_{15} \\ D_{21} & D_{22} & D_{23} & D_{24} & D_{25} \\ D_{31} & D_{32} & 0 & D_{34} & D_{35} \\ D_{41} & D_{42} & D_{43} & 0 & 0 \\ D_{51} & D_{52} & D_{53} & 0 & 0 \end{bmatrix}, \qquad (4.50)$$

and the spring matrix $\mathbf{C}(t)$

$$\mathbf{C}(t) = \Omega^2 \begin{bmatrix} 0 & 0 & C_{13} & C_{14} & C_{15} \\ 0 & 0 & C_{23} & C_{24} & C_{25} \\ 0 & 0 & C_{33} & 0 & 0 \\ 0 & 0 & 0 & C_{44} & 0 \\ 0 & 0 & 0 & 0 & C_{55} \end{bmatrix}. \qquad (4.51)$$

The populations of these matrices for the particular case are shown in Equations (4.49)–(4.51), the complete matrix entries are given in Appendix B. Apparently, neither the damping matrix nor the spring matrix are symmetric: this is not unusual for rotating systems, where the stiffness becomes variable with rotor speed [61], which is in this case given due to the hinge offset e. Furthermore, gyroscopic terms lead to an unsymmetric damping matrix [50].

4.2.5 Determination of reaction moments

To characterize the influence of the stabilizer bar on the helicopter, the restoring moments effected by the device need to be quantified. Therefore, the fuselage is virtually fixed and subjected to an imprinted rolling or pitching motion. Then, the respective restoring moments can be extracted from the model using Lagrangian multipliers.

The binding of the fuselage to the fixation point is denoted by $\vec{g}(\vec{q}, t)$. To express the following of the fuselage to the imprinted rolling and pitching motion of the fixation point, it can be stated that

$$\vec{g}(\vec{q}, t) = \begin{bmatrix} \phi(t) - \phi_{\text{imp}}(t) \\ \theta(t) - \theta_{\text{imp}}(t) \end{bmatrix} = \vec{0}, \qquad (4.52)$$

where $\phi_{\text{imp}}(t)$ and $\theta_{\text{imp}}(t)$ are the time trajectories of the imprinted rolling and pitching motion. In order to quantify the resulting roll and pitch moments in the fixation point, Equation (4.48) must be extended by a Lagrangian multiplier $\vec{\lambda}$ to yield

$$\mathbf{M} \cdot \ddot{\vec{q}} + \mathbf{D} \cdot \dot{\vec{q}} + \mathbf{C} \cdot \vec{q} - \vec{F}_{\text{gen}} + \mathbf{W} \cdot \vec{\lambda} = \vec{0}. \qquad (4.53)$$

Here, the matrix $\mathbf{W}$ is the transposed Jacobian of the binding $\vec{g}$ with respect to the generalized coordinates $\vec{q}$.

$$\mathbf{W} = \left(\frac{\partial \vec{g}}{\partial \vec{q}}\right)^T = \begin{bmatrix} 1 & 0 & 0 & 0 & 0 \\ 0 & 1 & 0 & 0 & 0 \end{bmatrix}^T. \qquad (4.54)$$

Then, the Lagrangian multiplier $\vec{\lambda}$ yields the reaction roll and pitch moments M_x and M_y:

$$\vec{\lambda} = \begin{bmatrix} M_x(t) \\ M_y(t) \end{bmatrix}. \qquad (4.55)$$

Since the Jacobian $\mathbf{W}$ cannot be inverted, Equation (4.53) first needs to be left multiplied with $\mathbf{W}^T \mathbf{M}^{-1}$ to yield

$$\mathbf{W}^T \ddot{\vec{q}} + \mathbf{W}^T \mathbf{M}^{-1} \mathbf{D} \dot{\vec{q}} + \mathbf{W}^T \mathbf{M}^{-1} \mathbf{C} \vec{q} - \mathbf{W}^T \mathbf{M}^{-1} \vec{F}_{\text{gen}} + \mathbf{W}^T \mathbf{M}^{-1} \mathbf{W} \vec{\lambda} = \vec{0}. \qquad (4.56)$$

Then, the square matrix $\mathbf{W}^T \mathbf{M}^{-1} \mathbf{W}$ can be inverted to solve for $\vec{\lambda}$:

$$\vec{\lambda} = \left(\mathbf{W}^T \mathbf{M}^{-1} \mathbf{W}\right)^{-1} \left(\mathbf{W}^T \ddot{\vec{q}} + \mathbf{W}^T \mathbf{M}^{-1} \mathbf{D} \dot{\vec{q}} + \mathbf{W}^T \mathbf{M}^{-1} \mathbf{C} \vec{q} - \mathbf{W}^T \mathbf{M}^{-1} \vec{F}_{\text{gen}}\right). \qquad (4.57)$$

With this, the simulation model is complete. Simulation results are shown in the following section.

4.3 Simulation results

This section shows simulation results for the previously introduced model. The results are separated for the motion simulation, which is used to find the following time of the stabilizer bar based on its flapping inertia and its rotational speed, and the force simulation, which is used to identify the optimal phase angle of the stabilizer bar.

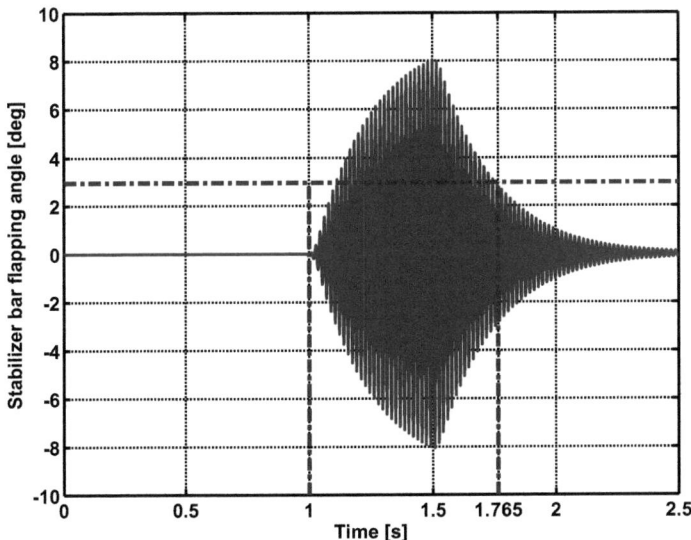

Figure 4.5: Motion simulation result for the stabilizer bar flapping angle.

4.3.1 Motion simulation

The motion simulations are used to find values for the stabilizer bar following time T_f as a function of rotor speed Ω and stabilizer bar flapping inertia I_{sb}. The following time is needed for the stabilizer module in the modular dynamic model in Chapter 5.

Figure 4.5 shows a simulation result of the stabilizer bar flapping angle for a rotor shaft ramp tilt of 20° at time $t = 1$ s. It shows the increase of the flapping angle to a maximum of approximately 8°, followed by an exponential decay. The time between the initiation of the tilt and the reaching of 37 % of the maximal amplitude accounts for 0.765 s, and is used as the stabilizer bar following time that is compared to the experimental results and can be used for the stabilizer bar module in the following chapter.

In the same fashion as in Figure 4.5, following times are found from simulations for different rotor speeds Ω and stabilizer bar flapping inertias I_{sb}. They are shown in Table 4.2. The simulation data shows that the following time depends strongly on the stabilizer bar flapping inertia I_{sb}, while the influence of the rotor speed Ω is rather small. In general, larger flapping inertias and rotor speeds lead to a larger following time. A comparison of the simulation results with experimental data is shown in Section 4.6.1.

Table 4.2: Simulated following times in [s] for different rotor speeds in [rad/s] and stabilizer inertias in [kgmm^2].

	$\Omega_1 = 270$	$\Omega_2 = 300$	$\Omega_3 = 330$	$\Omega_4 = 360$
$I_{sb,1} = 2.32$	0.580	0.580	0.578	0.575
$I_{sb,2} = 4.76$	0.640	0.645	0.647	0.649
$I_{sb,3} = 6.96$	0.693	0.694	0.695	0.698
$I_{sb,4} = 9.30$	0.745	0.748	0.754	0.765

4.3.2 Force simulation

For the force simulation, the previously introduced reaction moments are used to find the optimal stabilizer bar phase angle. The simulation model is run with different stabilizer bar phase angles between 0° and 90°, and the respective maxima of the moment M_y (disturbed axis) and moment M_x (undisturbed cross axis) are calculated. The results for these moments are shown in the top plot of Figure 4.6. The plot shows a relatively wide peak for the maximum of the restoring moment M_y at a phase angle of $\alpha = 62°$, and a more distinct minimum in the cross coupling moment M_x at a phase angle of $\alpha = 53°$. Basically, both criteria, either maximal restoring moment or minimal cross coupling moment, can be used to define the optimal stabilizer bar phase angle. By comparison of the time trajectories of the two moments, which are shown in the two plots at the bottom of Figure 4.6, it is, however, more expedient to choose the phase angle $\alpha = 53°$ as the optimal phase: its cross coupling moment is smaller than that for $\alpha = 62°$, while its restoring moment is almost equal. In Section 4.6.2, this simulation result is compared to experimental results.

4.4 Experimental setup

In order to have a comparison to the simulation results, an experimental setup is designed and built to collect measured data from the stabilizer bar. It is an improved version of the test rig presented in [84]. The core idea of the system is to mount a complete coaxial rotor setup with a passive lower rotor and an upper rotor augmented by a stabilizer bar on a six axis load cell. Thus the rotor forces and moments can be measured, and the influence of stabilizer bar design variations can be studied.

The test rig is designed such that the key design parameters of the stabilizer bar, namely flapping inertia I_{sb} and phase angle with respect to the blade pitching axis α can be varied. The rotors can be driven at different constant speeds, and for verification with the simulation the lower passive rotor can be fully shut down, leading to a single rotor test setup. The schematic layout of the setup is shown in Figure 4.7. The coaxial rotor system with stabilizer bar is held by a motor and gearbox housing, which is mounted on the load

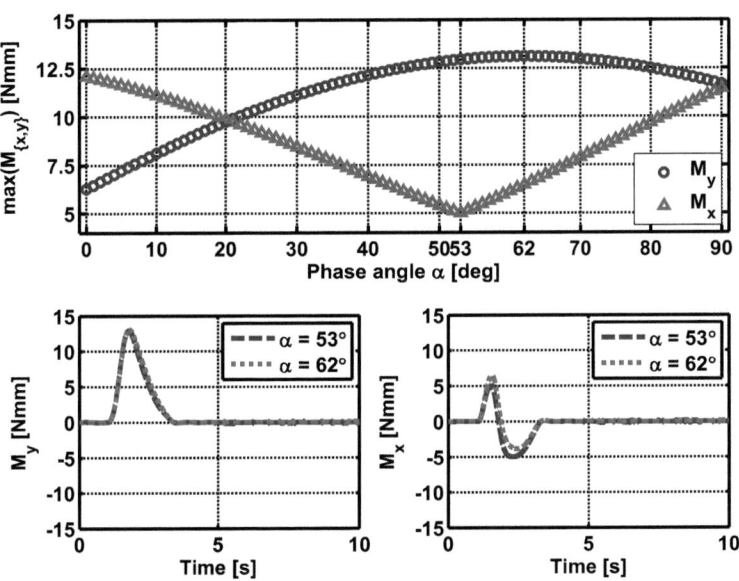

Figure 4.6: Force simulation result to identify the correct phase angle of the stabilizer bar for the muFly rotor system. The minimal cross coupling moment is found at 53°, while the maximal restoring moment is at 62°.

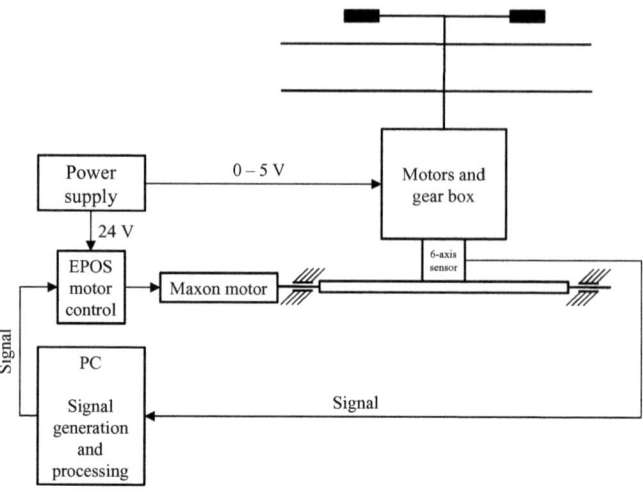

Figure 4.7: Schematic of the stabilizer bar test rig.

cells. The two Direct Current (DC) motors can be driven with up to 5 V from a lab power supply. The rotors and load cell are fixed on a metal plate, which can be tilted around one horizontal axis by another electrical motor. This motor is controlled with an EPOS motor controller, receiving its control input from a signal generation program running on the PC. The analog data from the load cell measurements is digitalized at a sampling rate of 10 KHz and sent to the computer for storage and digital signal processing. A photo of the test rig with its major components is shown in Figure 4.8. The tilting platform with the Maxon motor are supported by two roller bearings, which are mounted on a heavy aluminum structure and base plate. This construction minimizes the play and flexibility in the test rig and results in higher precision of the measurements.

4.5 Signal processing

Compared to a helicopter in free flight, the vibrations on a rigidly mounted set of rotors are even stronger. The high frequency vibrations from the rotor system are strong enough to cover the effect that is desired to be measured with the test rig. Figure 4.9 shows a typical unfiltered measurement as an example. At first glance, it can be seen that the signal contains a large portion of high frequency noise. The high frequency content of the signal becomes even more obvious, when the Short Time Fourier Transformation (STFT) [71] of the example signal is considered in Figure 4.10. Visible are the characteristic peak of the rotor frequency at 43 Hz, as well as the higher harmonics of the rotor at 86 Hz (second

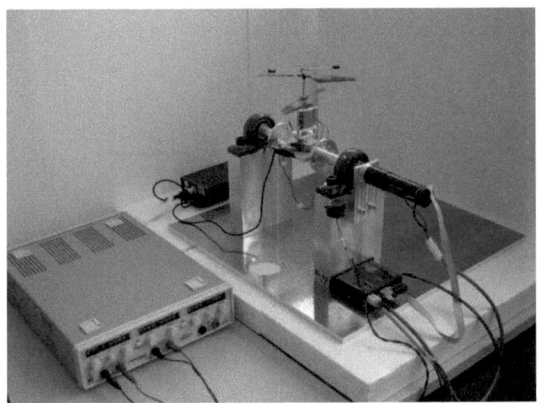

Figure 4.8: The test rig at the ASL.

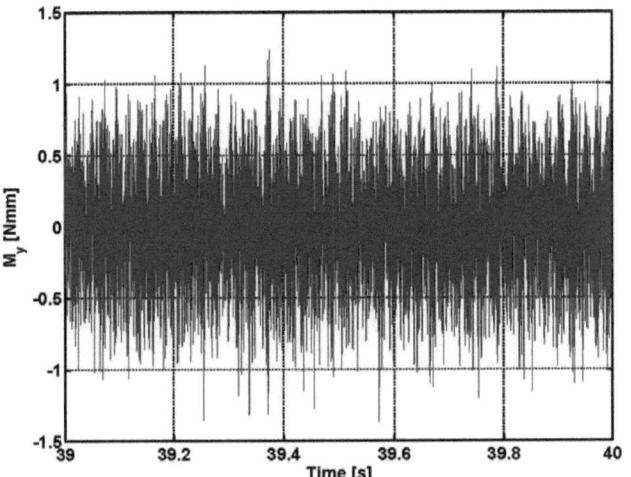

Figure 4.9: Unfiltered measurement of the moment M_y with running motor and no platform tilt.

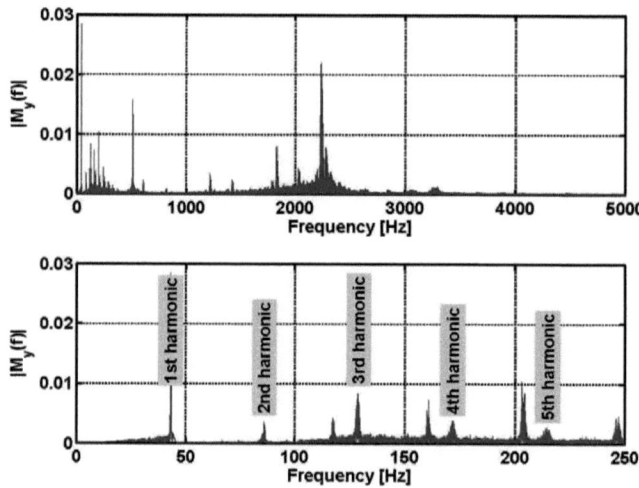

Figure 4.10: STFT of the example signal with a sampling rate of 10 KHz.

harmonic), at 129 Hz (third harmonic), at 172 Hz (fourth harmonic) and at 215 Hz (fifth harmonic). Another large peak is obvious in the range between 2 KHz and 3 KHz, which is a significantly higher frequency range than for any rotor higher harmonic or mechanical vibration in the test rig. The origin of this peak becomes clear when a measurement with no motors running is analyzed. Figure 4.11 shows that the peak in the Kilohertz range is prevalent, strongly indicating that this is a contribution of the test rig electronics, for instance the digital acquisition card of the PC, which digitalizes the load cell analog signals. In summary, digital filtering of the measured data is necessary to extract the data content relevant for the stabilizer bar characterization.

With a maximal platform tilting frequency of 0.25 Hz, the cutoff frequency for digital filtering can be safely placed at 5 Hz without risking the loss of relevant information. The digital filter used is a 2nd order Butterworth filter [71]. This filter is especially convenient for quantitative analysis, since it features a very sharp bend at the cutoff frequency, which keeps the pass band gain at 1 until very close to the cutoff frequency. Moreover, the attenuation of the filtered band is very high already shortly after passing the cut off frequency. The raw data is filtered with the filter described before, resulting in relatively smooth force and moment trajectories which can be used for quantitative analysis. in order to further improve the signal quality and to eliminate transients, sets of ten measurements are filtered and averaged afterwards. After this signal processing procedure, the data sets are ready for evaluation in the following section.

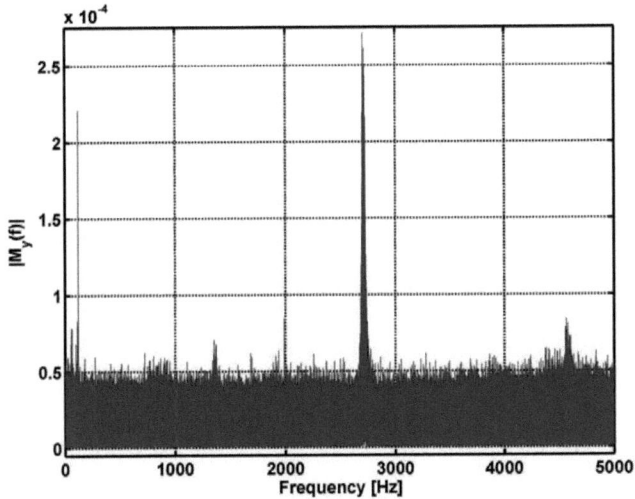

Figure 4.11: STFT of a measurement with no motors running.

4.6 Experimental results

In this section, post processed experimental results from the stabilizer bar test rig are shown. Two different measurements are performed: following time experiments, which characterize the stabilizer bar following time T_f as a function of its flapping inertia and rotational speed, and phase angle experiments, which are used to find the correct phase angle of the stabilizer bar for the muFly rotor system. The following time experiments are used to verify the moment simulation results, while the phase angle experiments are used to verify the force simulation results.

4.6.1 Following time experiments

For the following time experiments, the rotary gear on the test rig platform is subjected to a ramp tilt of 20° in 250 ms as shown in Figure 4.12. Moment trajectories of the disturbed axis are shown in Figure 4.13 for a rotor speed of 270 rad/s, in Figure 4.14 for a rotor speed of 300 rad/s, in Figure 4.15 for a rotor speed of 330 rad/s, and in Figure 4.16 for a rotor speed of 360 rad/s. Each figure shows the trajectories for four different flapping inertias I_{sb}.

The trajectories show similar behavior of the stabilizer bar with two characteristic peaks at acceleration and deceleration of the platform. All trajectories converge to the final static moment of 11.2 Nmm, however at different speeds. The respective following times are measured at 63 % of the final static moment, which is at 7.1 Nmm. The following times

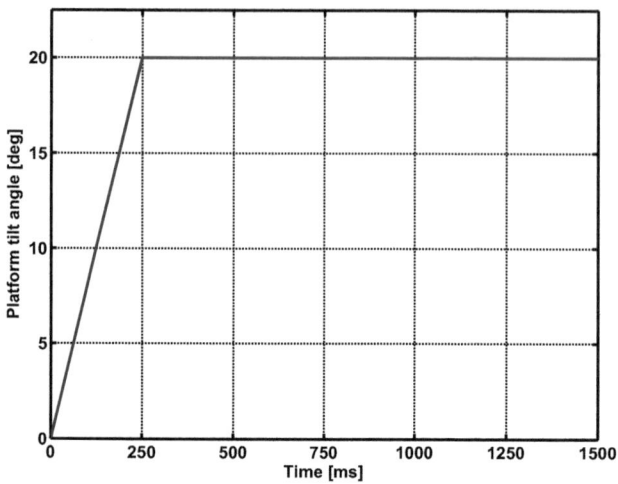

Figure 4.12: Platform tilt of the stabilizer bar test rig for the following time experiments.

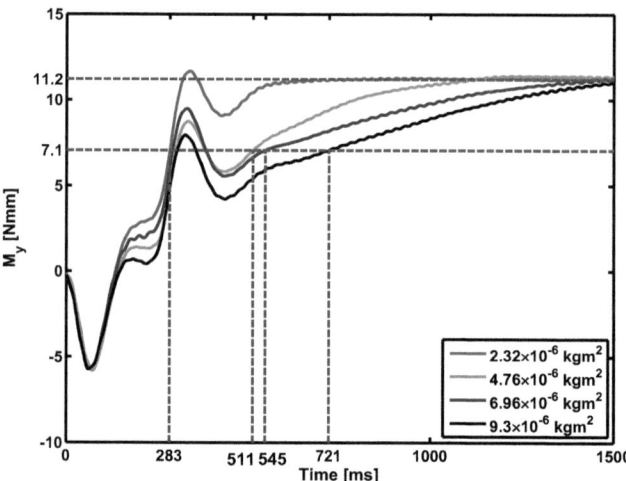

Figure 4.13: Moment trajectories and following times for four different stabilizer bar inertias at a rotor speed of 270 rad/s.

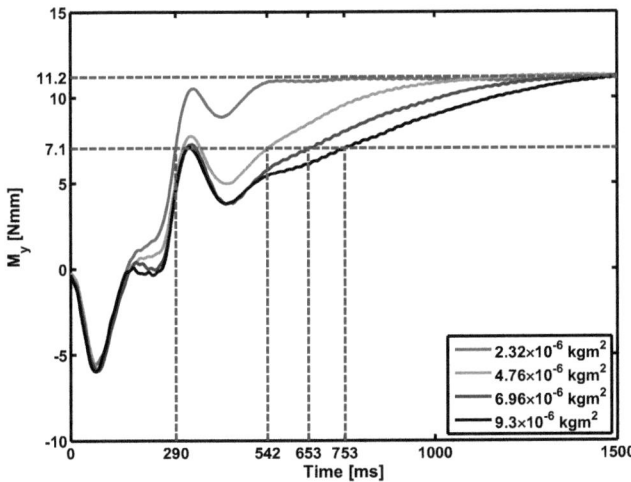

Figure 4.14: Moment trajectories and following times for four different stabilizer bar inertias at a rotor speed of 300 rad/s.

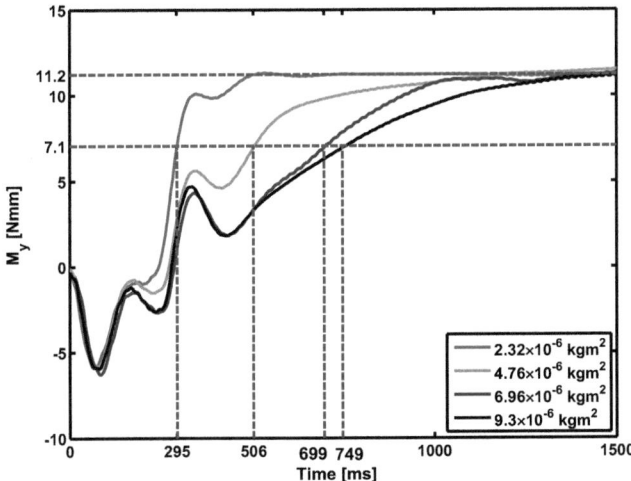

Figure 4.15: Moment trajectories and following times for four different stabilizer bar inertias at a rotor speed of 330 rad/s.

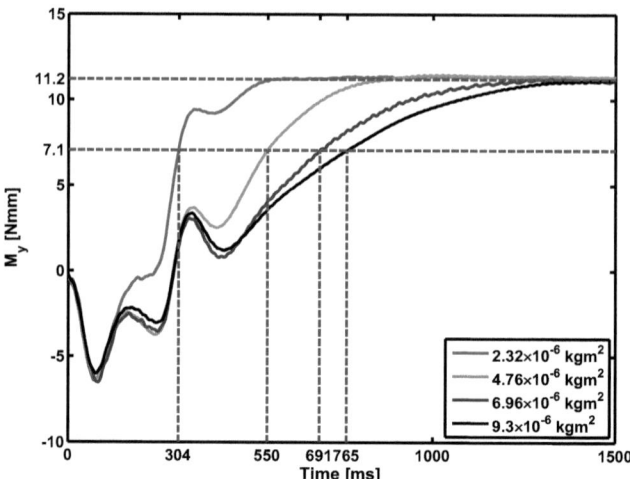

Figure 4.16: Moment trajectories and following times for four different stabilizer bar inertias at a rotor speed of 360 rad/s.

are summarized in Table 4.3 according to their stabilizer flapping inertias $I_{sb,i=\{1,2,3,4\}}$ and rotor speeds $\Omega_{i=\{1,2,3,4\}}$. By inspection of the following times shown in the table, it becomes obvious that they are much more sensitive to variations of the stabilizer flapping inertia than to variations of the rotor speed, which corresponds to the observations made for the simulation results.

To compare simulation and experiment, the data from Table 4.2 and from Table 4.3 is plotted in Figure 4.17 for the four different rotor speeds $\Omega_{i=\{1,2,3,4\}}$. The comparison shows a relatively good match between simulation and experiment for higher values of the stabilizer bar inertia, while for the lowest value of the flapping inertia there is a discrepancy of approximately 300 ms. A possible explanation is the short measuring time for such small

Table 4.3: Experimental following times in [s] for different rotor speeds in [rad/s] and stabilizer inertias in [kgmm²].

	$\Omega_1 = 270$	$\Omega_2 = 300$	$\Omega_3 = 330$	$\Omega_4 = 360$
$I_{sb,1} = 2.32$	0.283	0.290	0.295	0.304
$I_{sb,2} = 4.76$	0.511	0.542	0.506	0.550
$I_{sb,3} = 6.96$	0.545	0.653	0.699	0.691
$I_{sb,4} = 9.30$	0.721	0.753	0.749	0.765

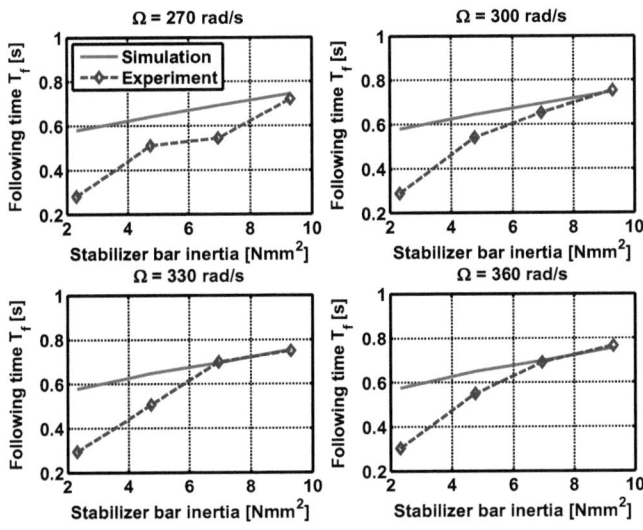

Figure 4.17: Comparison between simulation and experiments at different rotor speeds.

values of the stabilizer bar inertia: the measurement is taken, while the platform is still in motion, which can lead to biasing by the inertial moment due to acceleration and deceleration of the platform.

4.6.2 Phase angle experiments

For the phase angle experiments, the rotary gear on the test rig platform is subjected to a sinusoidal tilt of 20° amplitude at a frequency of 0.25 Hz. The corresponding input signal is shown in Figure 4.18. In order to obtain meaningful results for an optimal phase angle, the resulting moment maxima on the disturbed axis and the undisturbed cross axis need to be considered. A maximal restoring moment on the disturbed axis, or a minimal cross coupling moment on the undisturbed axis are both indicators for an optimal phase angle. The results for the moment maxima on both axes are shown in Figure 4.19. Obviously, the results for the disturbed axis in the top figure are not conclusive, since the moment stays relatively constant with varied phase angle. Apparently, the maximal restoring moment is smaller than other moments measured by the sensor, leading to the almost constant values over the different phase angles. The results for the undisturbed cross axis in the bottom of the figure, however, show a significant minimum in the range between 60° and 80° for all rotor speeds from 270 rad/s to 330 rad/s, with the absolute minimum at 60°. This result is conclusive, since a minimal moment on the cross axis indicates optimal phasing of the

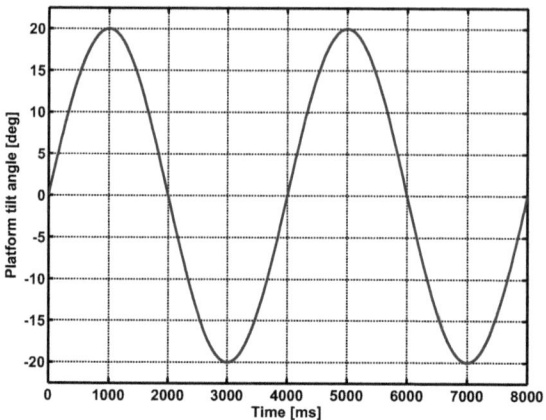

Figure 4.18: Platform tilt of the stabilizer bar test rig for the phase angle experiments.

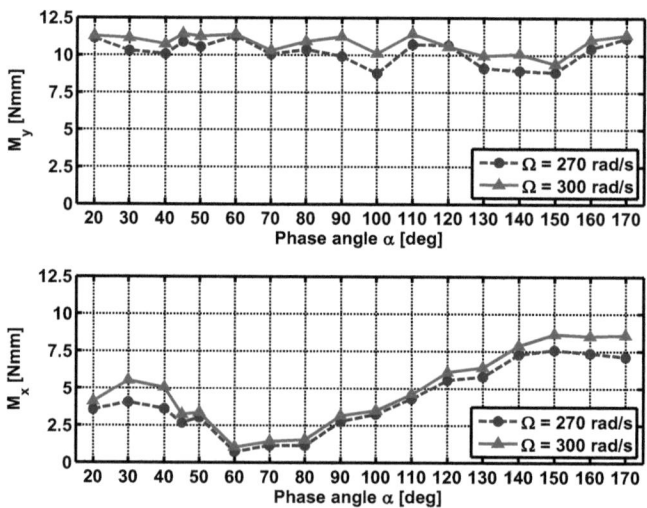

Figure 4.19: Moment as a function of phase angle for the disturbed axis (top) and the undisturbed cross axis (bottom).

stabilizer bar. It is also in relatively good agreement with the simulation result of 53° for the minimal moment on the undisturbed axis.

4.7 Summary

In this chapter, an analytical simulation model of the stabilizer bar is developed and compared to experimental results. The stabilizer bar and flapping blades are treated as a multi-body system, for which the equations of motion are derived following the Euler-Lagrange method. In order to allow for quantification of steering moments, the equations of motion are extended by Lagrangian multipliers to calculate the reaction moments in the support of the system. Two types of simulations are performed with the model: motion simulations to obtain the following time of the stabilizer bar, and force simulations, which allow for the identification of the correct stabilizer phase angle with respect to the blade pitching axis.

For the experimental validation, a test rig is developed, which allows for tilt angle disturbances in one axis and measurement of the resulting forces and moments in six axes. With this device, the following time and phase angle for the muFly helicopter can be determined. The experimental procedure requires different input signals for the following time and phase angle measurements, a high number of repetitions of the measurements to increase the reliability of the data, and signal processing of the sensor data to eliminate high frequency noise stemming from the electronics and the rotors.

In agreement with the simulation results, the experimental results for the following time show an increased following time with increased rotor speed and flapping inertia. The flapping inertia has, however, a significantly larger influence on the following time than the rotor speed. For the stabilizer bar mounted on the muFly helicopter, the following time accounts for 765 ms. On the other hand, the optimal phase angle from simulation is at 53°, where a minimum in the undesired cross axis moment occurs. The same minimum can be found in the experimental results at a phase angle of 60° for the muFly rotor system, which is still in relatively good agreement. The following time and known phase angle are important results for the stabilizer bar module in the following chapter.

In general, the stabilizer bar is a useful device for passive stabilization of small RC helicopters. However, as soon as the main concern becomes helicopter autonomy, the stabilizer bar should be removed from the system to save for mass and drag and thus increase the total flight time. Then, the micro helicopter needs to be controlled actively via closed loop controllers, which is certainly feasible nowadays. Another possibility for passive stabilization without a stabilizer bar is shown in Chapter 6 of this work.

Chapter 5

Modular dynamic model

As the literature review in Chapter 1 has shown, most dynamic simulation models for UAVs and MAVs are developed for existing systems. They are used as identification models, which are necessary for model based controller synthesis. However, these identification models rely on an existing helicopter for parameter identification, and oftentimes several physical subsystems are summarized in one parameter, for example a time constant for the upper rotor in connection with a stabilizer bar. Hence, such models are unsuitable to simulate dynamic helicopter behavior in the design phase, where it is desirable to assess different setups of the helicopter without having the physical prototype at hand.

The goal of this chapter is to introduce a modular dynamic model, which is developed to fill this gap and quantitatively support design decisions by giving insight into several fields: the blade flapping dynamics of the coaxial rotor configuration, the effect of the stabilizer bar, and two different steering principles for micro helicopters. The underlying equations of each module are derived, and simulation results are shown for different variants of a coaxial micro helicopter.

Hence, this chapter is organized as follows: in Section 5.1, the structure of the model and its different modules are introduced. Subsequently, the underlying equations for each module are derived in Section 5.2 for the hingeless rotor module, in Section 5.3 for the stabilizer bar module, in Section 5.4 for the swash plate steering module and in Section 5.5 for the center of mass displacement steering module. Simulation results are shown in Section 5.6. The contents of this chapter are an elaboration and extension of the model presented in [18].

5.1 Model structure

The dynamic model consists of several modules, which allows for different combinations to simulate different setups of coaxial micro helicopters. These modules are:

- A rotor module, which allows for simulation of rotors with different parametrizations ranging from teetering hinged to hingeless rotors. The module also takes into account passively excited flapping due to roll, pitch and forward velocities of the helicopter.

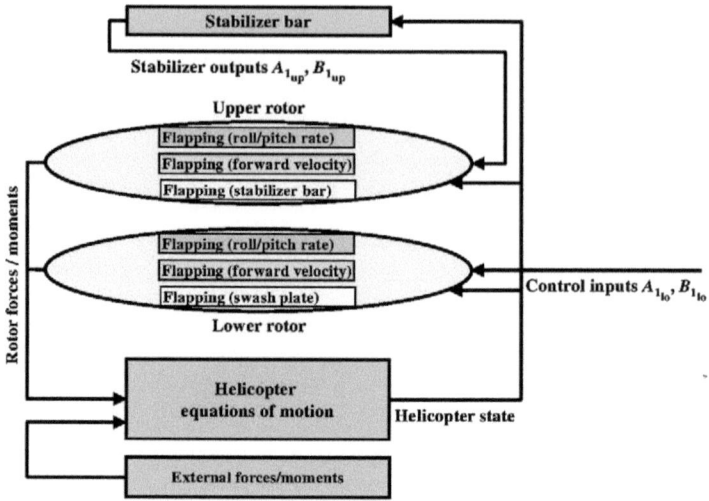

Figure 5.1: Model structure for swash plate steering with stabilizer bar.

- A stabilizer bar module, which can be added to the upper rotor to simulate the influence of the passive stabilization device.
- A swash plate steering module, which can be added to the lower rotor of the coaxial helicopter.
- A center of mass displacement steering module, which can be added instead of the swash plate steering module to simulate an alternative steering option for micro helicopters.

The details of the modules and the derivation of their governing equations will be shown in the following sections.

Figure 5.1 shows a setup of the model for conventional swash plate steering including stabilizer bar dynamics. The helicopter equations of motion correspond to Equation (3.16). The resulting helicopter state is fed into the lower and upper rotor modules to calculate the passive flapping contributions due to roll, pitch and forward velocities. Furthermore, the helicopter state is fed into the stabilizer bar module to calculate its output, which serves as a cyclic pitch input to the upper rotor. Similarly, the swash plate mechanism transmits a cyclic pitch input on the lower rotor, however, governed by two active control inputs $A_{1_{lo}}$ and $B_{1_{lo}}$. Both cyclic pitch inputs lead to the third flapping contribution in their respective rotor. Based on the resultant total flapping angles according to Equation (3.13), the rotor forces and moments are calculated for each rotor and fed back to the equations of motion. All other forces and moments acting on the helicopter are introduced via an external forces

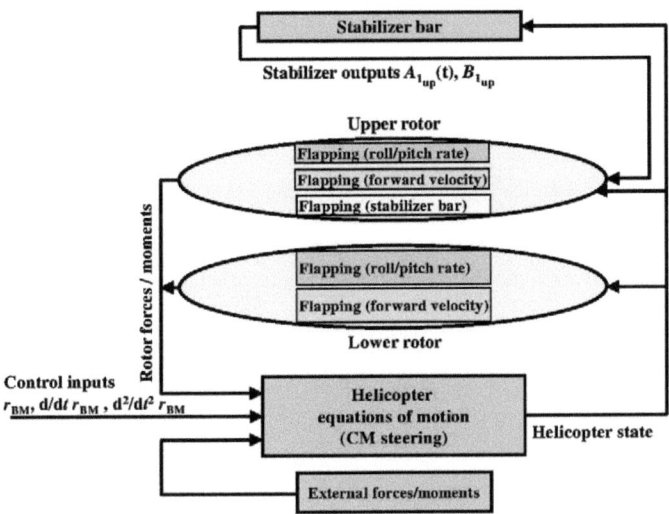

Figure 5.2: Model structure for center of mass displacement steering with stabilizer bar.

and moments block.

A different setup for the coaxial micro helicopter is shown in Figure 5.2. Here, the swash plate steering has been replaced by center of mass displacement steering. As a result, the lower rotor does not produce blade flapping due to swash plate action anymore. Instead, the helicopter equations of motion are modified such that they accept the control inputs $\vec{r}_{BM}$, $\dot{\vec{r}}_{BM}$ and $\ddot{\vec{r}}_{BM}$, with $\vec{r}_{BM}$ being the vector from the center of mass of the helicopter to its moveable mass.

All modules of the simulation model are implemented in Matlab/Simulink, as shown in Figure 5.3. The stabilizer bar module, the different steering principles and other physical effects like fuselage drag, rotor shaft gyroscopic moments and disturbance forces and moments can be switched on and off via selector switches.

Principally, it is possible to deactivate the stabilizer bar module, which deactivates the flapping contribution due to cyclic pitch input on the upper rotor. This is an option that will be investigated more in depth in Chapter 6. Due to their passive flapping contributions, the upper and lower rotor act as feedback loops on the equations of motion, which could be exploited for passive roll and pitch stabilization without a stabilizer bar.

In the following sections, the different modules are introduced in more detail.

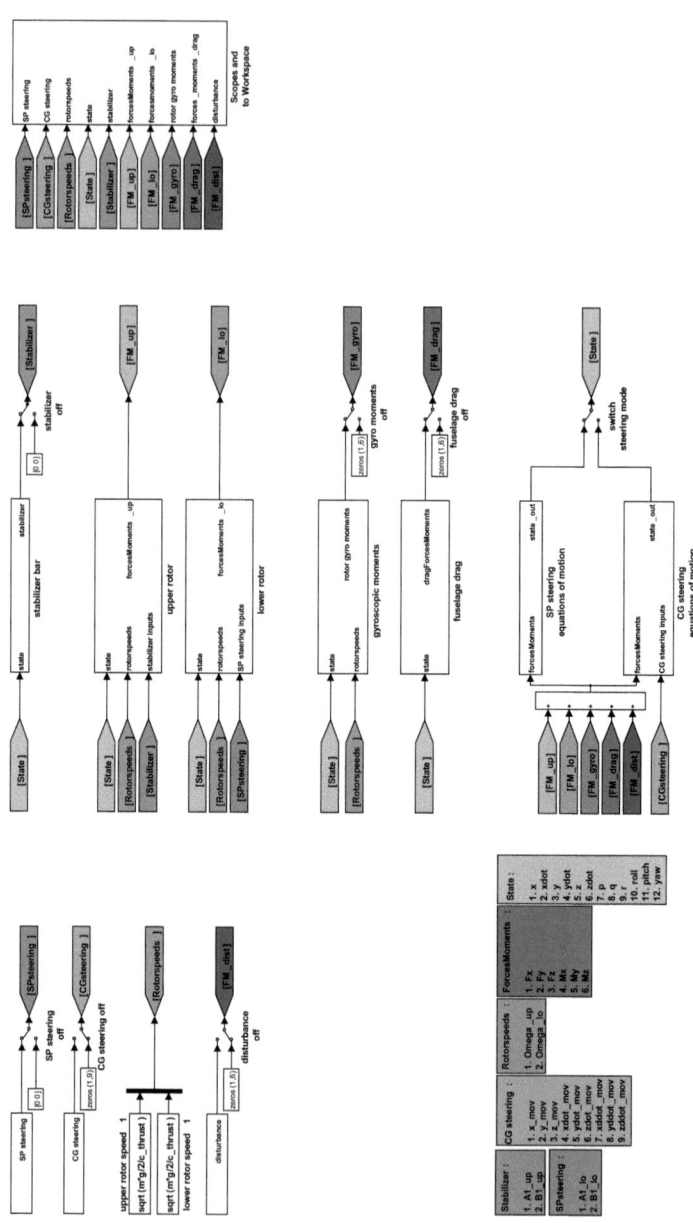

Figure 5.3: Simulink implementation of the modular simulation model.

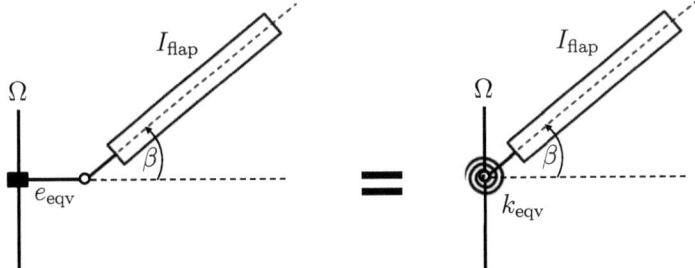

Figure 5.4: Equivalence of hinge offset and flapping spring stiffness at equal rotor speed Ω.

5.2 Hingeless rotor module

In this section, the dynamic equations for a coaxial hingeless rotor system are derived. This includes the derivation of a substitute model for the flexible rotor blade, the calculation of resulting rotor forces and moments and the derivation of the passive blade flapping angles due to roll, pitch and forward velocities of the helicopter.

5.2.1 Basic model and parameter equivalence

The modeling of the blade flapping and the subsequent calculation of the resulting steering moments is based on a hingeless rotor, i.e. a rotor that has a flexible connection instead of a hinge between the blade root and the rotor shaft. To approximate the hingeless rotor for simplification of the model, it is substituted by a hinged rotor with hinge offset, i.e. a vertical distance between rotor shaft and blade root, and a torsional spring in the flapping hinge. This simplification can be found for instance in [21] and is shown in Figure 3.6. The blade is attached to the shaft rotating at the speed Ω through a hinge with hinge offset e and torsional stiffness k_flap. The blade itself has a flapping inertia I_flap.

For this substitute model it can be shown (for instance in [76]) that hinge offset e and torsional stiffness k_flap are equivalent and can be transformed into each other for a fixed rotor speed Ω. In order to show this, first of all the moment equilibrium around the flapping hinge is formulated for a rotor with hinge offset but without spring stiffness (see Figure 5.4 (left)):

$$\Omega^2 \beta \left(I_\text{flap} + \frac{e M_\text{bld}}{g} \right) = 0. \tag{5.1}$$

The static blade moment M_bld is defined in Equation (3.11).

Similarly, the moment equilibrium around the flapping hinge is formulated for a rotor without hinge offset but with torsional stiffness in the hinge (see Figure 5.4 (right)):

$$\Omega^2 \beta I_\text{flap} + \beta k_\text{flap} = 0. \tag{5.2}$$

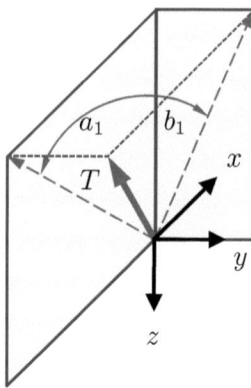

Figure 5.5: Projection of the rotor thrust T on the body-fixed axes with the longitudinal and lateral flapping angles a_1 and b_1.

Equating the two equilibria, the result can either be solved for an equivalent hinge offset e_{eqv} or for an equivalent flapping stiffness k_{eqv}:

$$e_{\text{eqv}} = \frac{g}{\Omega^2 M_{\text{bld}}} k_{\text{flap}}, \tag{5.3}$$

$$k_{\text{eqv}} = \frac{\Omega^2 M_{\text{bld}}}{g} e. \tag{5.4}$$

Hence, either hinge offset e or flapping stiffness k_{flap} can be eliminated by inclusion into the respective other parameter, which makes the following derivations significantly easier. They will be made in terms of a combined hinge offset e_{com}, which is the superposition of the actual hinge offset e and the equivalent hinge offset e_{eqv} that replaces the flapping spring stiffness k_{flap}:

$$e_{\text{com}} = e + e_{\text{eqv}}. \tag{5.5}$$

All the following derivations use this combined parameter.

5.2.2 Derivation of the rotor forces and moments

Rotor blade flapping leads to a tilting of the respective rotor tip path plane and therefore to a misalignment of the rotor disc and the helicopter fuselage. Therefore, the rotor thrust force T, which is always oriented perpendicular to the tip path plane, has to be projected on the body-fixed coordinate axes. This situation is shown in Figure 5.5. For known thrust forces of the upper and lower rotor $T_{\{\text{up,lo}\}}$, the resulting forces projected on the body-fixed coordinate axes can be derived as:

$$F_{x_{\{up,lo\}}} = T_{\{up,lo\}} \left[-\cos\left(b_{1_{\{up,lo\}}}\right) \sin\left(a_{1_{\{up,lo\}}}\right)\right], \tag{5.6}$$

$$F_{y_{\{up,lo\}}} = T_{\{up,lo\}} \left[\sin\left(b_{1_{\{up,lo\}}}\right)\right], \tag{5.7}$$

$$F_{z_{\{up,lo\}}} = T_{\{up,lo\}} \left[-\cos\left(b_{1_{\{up,lo\}}}\right) \cos\left(a_{1_{\{up,lo\}}}\right)\right]. \tag{5.8}$$

As a result of the tilting of the tip path plane with respect to the fuselage, the thrust force vector does not pass through the center of mass of the helicopter any more, which results in a moment. Furthermore, the spring stiffness in each blade, now denoted by e_{com}, is loaded, which leads to a spring moment. For the three body axes, these moments are:

$$M_{x_{\{up,lo\}}} = b_{1_{\{up,lo\}}} \left[\frac{1}{2}\frac{e_{\text{com}}}{R} m_{\text{bld}} \left(\Omega_{\{up,lo\}} R\right)^2\right] + F_{y_{\{up,lo\}}} d_{\{up,lo\}}, \tag{5.9}$$

$$M_{y_{\{up,lo\}}} = a_{1_{\{up,lo\}}} \left[\frac{1}{2}\frac{e_{\text{com}}}{R} m_{\text{bld}} \left(\Omega_{\{up,lo\}} R\right)^2\right] - F_{x_{\{up,lo\}}} d_{\{up,lo\}}, \tag{5.10}$$

$$M_{z_{\{up,lo\}}} = 0. \tag{5.11}$$

The moments in Equations (5.9)–(5.11) do only take into account the rotor moments due to flapping. Therefore, the z-moment in Equation (5.11) is equal to zero. However, the difference of the drag torques of the upper and lower rotor constitutes the total moment acting on the z-axis of the helicopter:

$$M_z = T_{\text{up}}^{\text{drag}} - T_{\text{lo}}^{\text{drag}}. \tag{5.12}$$

The thrusts $T_{\{up,lo\}}$ and drag torques $T_{\{up,lo\}}^{\text{drag}}$ of the upper and lower rotor are calculated from a simplified aerodynamic model, assuming them to be proportional to the squared rotor speeds:

$$T_{\text{up}} = k_{\text{up}}^{\text{lift}} \Omega_{\text{up}}^2 \qquad\qquad T_{\text{up}}^{\text{drag}} = k_{\text{up}}^{\text{drag}} \Omega_{\text{up}}^2, \tag{5.13}$$

$$T_{\text{lo}} = k_{\text{lo}}^{\text{lift}} \Omega_{\text{lo}}^2 \qquad\qquad T_{\text{lo}}^{\text{drag}} = k_{\text{lo}}^{\text{drag}} \Omega_{\text{lo}}^2. \tag{5.14}$$

5.2.3 Derivation of the blade flapping angles

In the following, the governing equations for flapping due to roll, pitch and forward velocities of the helicopter, i.e. the passive flapping contributions, are derived. For this, the same assumptions as in [77] are made:

(a) Aerodynamic forces are considered to act from the hinge to the tip.
(b) The reverse flow region is ignored.
(c) The airfoil lift characteristics are linear and free of stall and compressibility effects.
(d) The blade motion consists of only coning and first harmonic flapping.
(e) Small angle assumptions are valid.

Following the same strategy as in [77], the flapping angles are derived for the upper and the lower rotor based on moment equilibria around the flapping hinges of the blades.

Flapping due to roll and pitch velocity

When the helicopter performs rolling or pitching motions, the individual rotor blade is subjected to three types of moments. These are the centrifugal moment M^{cf}, the aerodynamic moment M^{aero} and the gyroscopic moment M^{gyro}. By making use of the combined hinge offset e_{com}, the contribution of the torsional spring loading is already included in the centrifugal moment and does not need to be introduced as an additional moment. Following the sign convention defined in Figure 3.4, these moments can be formulated for the upper and lower rotor, which turn in opposite directions. Starting with the upper rotor, the three moments are:

$$M^{\text{cf}}_{\text{up}} = \Omega^2_{\text{up}} \frac{e_{\text{com}} M_{\text{bld}}}{g} \left(b^{\text{rot}}_{1\text{up}} \sin \psi_{\text{up}} + a^{\text{rot}}_{1\text{up}} \cos \psi_{\text{up}} \right), \tag{5.15}$$

$$M^{\text{aero}}_{\text{up}} = \frac{\rho}{8ac} (\Omega_{\text{up}} R)^2 \left(1 - \frac{e_{\text{com}}}{R}\right)^2 R^2 \left[\left[\frac{p}{\Omega_{\text{up}}} - a^{\text{rot}}_{1\text{up}}\left(1 - \frac{\mu^2}{2}\right)\right] \sin \psi_{\text{up}} \right.$$
$$\left. + \left[\frac{q}{\Omega_{\text{up}}} + b^{\text{rot}}_{1\text{up}}\left(1 + \frac{\mu^2}{2}\right)\right] \cos \psi_{\text{up}} \right], \tag{5.16}$$

$$M^{\text{gyro}}_{\text{up}} = 2\Omega_{\text{up}} I_{\text{flap}} \left(-q \sin \psi_{\text{up}} + p \cos \psi_{\text{up}}\right). \tag{5.17}$$

These moments have to equilibrate around the flapping hinge, such that the following holds:

$$M^{\text{cf}}_{\text{up}} + M^{\text{aero}}_{\text{up}} + M^{\text{gyro}}_{\text{up}} = 0. \tag{5.18}$$

Insertion of the moments into Equation (5.18), separation of the sine and cosine terms and simultaneous solution of the resulting equations for the flapping angles $a^{\text{rot}}_{1,\text{up}}$ and $b^{\text{rot}}_{1,\text{up}}$ yields the expressions

$$a^{\text{rot}}_{1\text{up}} = \frac{\frac{-\frac{16}{\gamma}\left(\frac{q}{\Omega_{\text{up}}}\right)}{\left(1-\frac{e_{\text{com}}}{R}\right)^2} + \left(\frac{p}{\Omega_{\text{up}}}\right)}{1 - \frac{\mu^2}{2}} + \frac{\frac{12}{\gamma}\frac{e_{\text{com}}}{R}}{\left(1-\frac{e_{\text{com}}}{R}\right)^3} \left[\frac{-\frac{16}{\gamma}\left(\frac{p}{\Omega_{\text{up}}}\right)}{\left(1-\frac{e_{\text{com}}}{R}\right)^2} - \left(\frac{q}{\Omega_{\text{up}}}\right)\right]}{1 - \frac{\mu^4}{4}}, \tag{5.19}$$

$$b^{\text{rot}}_{1\text{up}} = \frac{\frac{-\frac{16}{\gamma}\left(\frac{p}{\Omega_{\text{up}}}\right)}{\left(1-\frac{e_{\text{com}}}{R}\right)^2} - \left(\frac{q}{\Omega_{\text{up}}}\right)}{1 + \frac{\mu^2}{2}} + \frac{\frac{12}{\gamma}\frac{e_{\text{com}}}{R}}{\left(1-\frac{e_{\text{com}}}{R}\right)^3} \left[\frac{\frac{16}{\gamma}\left(\frac{q}{\Omega_{\text{up}}}\right)}{\left(1-\frac{e_{\text{com}}}{R}\right)^2} - \left(\frac{p}{\Omega_{\text{up}}}\right)\right]}{1 - \frac{\mu^4}{4}}. \tag{5.20}$$

Similarly, the three moments can be formulated for the lower rotor according to the sign convention in Figure 3.4, which accounts for several sign changes in the otherwise identical expressions.

$$M^{\text{cf}}_{\text{lo}} = \Omega^2_{\text{lo}} e_{\text{com}} \frac{M_{\text{bld}}}{g} \left(b^{\text{rot}}_{1\text{lo}} \sin \psi_{\text{lo}} - a^{\text{rot}}_{1\text{lo}} \cos \psi_{\text{lo}} \right), \tag{5.21}$$

$$M^{\text{aero}}_{\text{lo}} = \frac{\rho}{8ac} (\Omega_{\text{lo}} R)^2 \left(1 - \frac{e_{\text{com}}}{R}\right)^2 R^2 \left[\left[-\frac{p}{\Omega_{\text{lo}}} + a^{\text{rot}}_{1\text{lo}}\left(1 - \frac{\mu^2}{2}\right)\right] \sin \psi_{\text{lo}} \right.$$
$$\left. + \left[-\frac{q}{\Omega_{\text{lo}}} + b^{\text{rot}}_{1\text{lo}}\left(1 + \frac{\mu^2}{2}\right)\right] \cos \psi_{\text{lo}} \right], \tag{5.22}$$

$$M^{\text{gyro}}_{\text{lo}} = 2\Omega_{\text{lo}} I_{\text{flap}} \left(q \sin \psi_{\text{lo}} - p \cos \psi_{\text{lo}}\right), \tag{5.23}$$

Again, the moments must equilibrate around the flapping hinge to yield

$$M_{\text{lo}}^{\text{cf}} + M_{\text{lo}}^{\text{aero}} + M_{\text{lo}}^{\text{gyro}} = 0. \tag{5.24}$$

Insertion of the moments into Equation (5.24), separation of the sine and cosine terms and simultaneous solution for the flapping angles $a_{1_{\text{lo}}}^{\text{rot}}$ and $b_{1_{\text{lo}}}^{\text{rot}}$ leads to the expressions

$$a_{1_{\text{lo}}}^{\text{rot}} = \frac{\frac{-\frac{16}{\gamma}\left(\frac{q}{\Omega_{\text{lo}}}\right)}{\left(1-\frac{e_{\text{com}}}{R}\right)^2} - \left(\frac{p}{\Omega_{\text{lo}}}\right)}{1-\frac{\mu^2}{2}} + \frac{\frac{\frac{12}{\gamma}\frac{e_{\text{com}}}{R}}{\left(1-\frac{e_{\text{com}}}{R}\right)^3}\left[\frac{\frac{16}{\gamma}\left(\frac{p}{\Omega_{\text{lo}}}\right)}{\left(1-\frac{e_{\text{com}}}{R}\right)^2} - \left(\frac{q}{\Omega_{\text{lo}}}\right)\right]}{1-\frac{\mu^4}{4}}, \tag{5.25}$$

$$b_{1_{\text{lo}}}^{\text{rot}} = \frac{\frac{-\frac{16}{\gamma}\left(\frac{p}{\Omega_{\text{lo}}}\right)}{\left(1-\frac{e_{\text{com}}}{R}\right)^2} + \left(\frac{q}{\Omega_{\text{lo}}}\right)}{1+\frac{\mu^2}{2}} + \frac{\frac{\frac{12}{\gamma}\frac{e_{\text{com}}}{R}}{\left(1-\frac{e_{\text{com}}}{R}\right)^3}\left[\frac{-\frac{16}{\gamma}\left(\frac{q}{\Omega_{\text{lo}}}\right)}{\left(1-\frac{e_{\text{com}}}{R}\right)^2} - \left(\frac{p}{\Omega_{\text{lo}}}\right)\right]}{1-\frac{\mu^4}{4}}, \tag{5.26}$$

that have the same structure as in Equations (5.19)–(5.20), but different signs for the terms describing the axis perpendicular to the respective flapping axis (cross-direction of rotation).

Simulation of these expressions shows why the coaxial rotor configuration is principally more stable in roll and pitch than, for instance, the conventional rotor configuration. This is shown in the result plots of Figures 5.6–5.9. Each of the result plots displays the horizontal rotor forces and moments F_x, F_y, M_x and M_y of the upper rotor (solid blue), the lower rotor (dashed red) and their resultants (dashed green) in the helicopter fixed coordinate frame. Figure 5.6 shows the rotor forces and moments for a roll moment disturbance of 1 Nmm of a rotor without hinge offset. While the y-forces and the respective x-moments of the rotors counteract the disturbance with amplitudes in the same direction, the x-forces and respective y-moments react in opposite directions, canceling each other out. Since the moments are achieved by thrust vector reorientation only (no hinge offset to add rotor stiffness), the moments can not cancel out perfectly due to the different distances of the rotors from the center of mass of the helicopter. In Figure 5.7, the same result is shown for a flapping hinge offset of 10 mm. Qualitatively, the same result can be observed. While the cross-direction contributions cancel out, the contributions on the disturbed axis counteract the disturbance. In contrast to the rotor without hinge offset, however, the magnitude of the restoring moment is approximately ten times larger, and also the forces and moments build up much more rapidly than in the case without hinge offset. This corresponds to the general understanding that helicopters with stiff rotor heads are much more maneuverable due to their faster buildup of steering moments [21]. Figures 5.8–5.9 show the same results for a pitch disturbance. There, the restoring force and moment resultants are F_x and M_y, while F_y and M_x cancel out almost completely. Compared to a conventional single main rotor, the coaxial rotor results in stronger moment disturbance rejection (two rotors yield a restoring moment instead of one), while the undesired cross terms are canceled out as a result of the opposite directions of rotation of the rotors. Furthermore, introduction of a

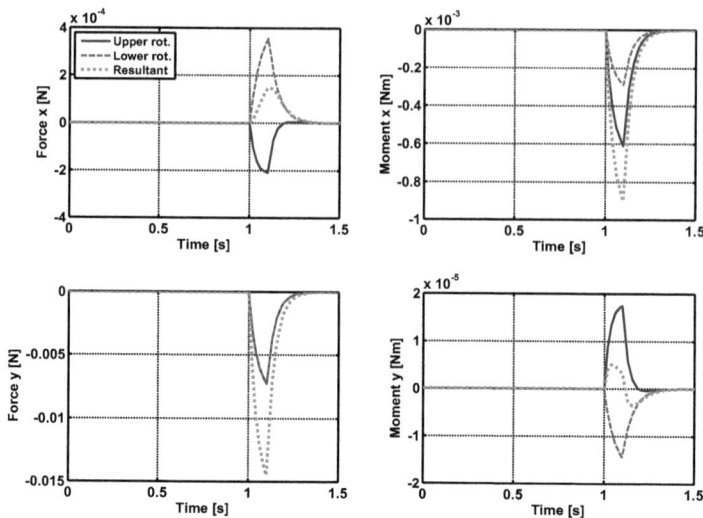

Figure 5.6: Horizontal rotor forces and moments for a roll moment disturbance of 1 Nmm without hinge offset.

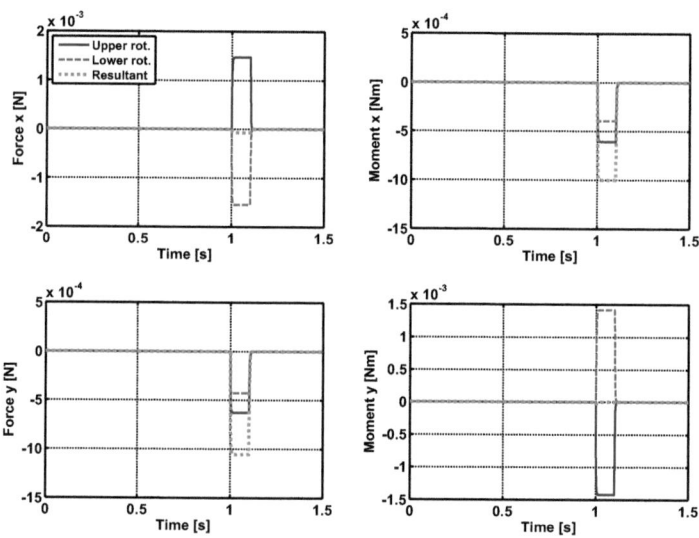

Figure 5.7: Horizontal rotor forces and moments for a roll moment disturbance of 1 Nmm with a hinge offset of 10 mm.

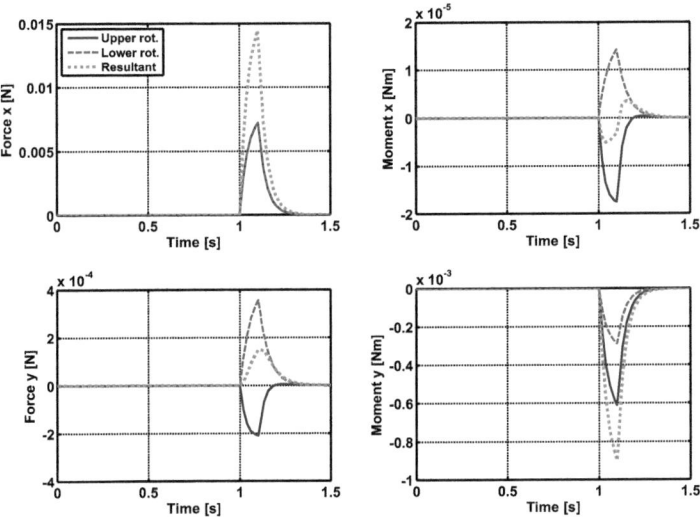

Figure 5.8: Horizontal rotor forces and moments for a pitch moment disturbance of 1 Nmm without hinge offset.

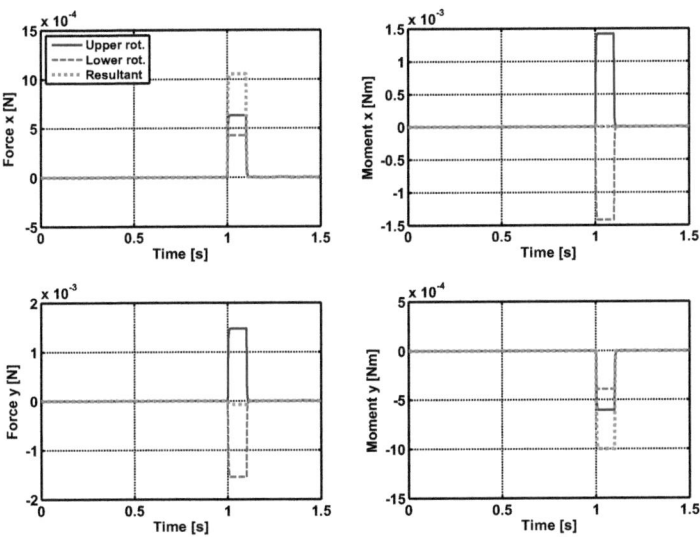

Figure 5.9: Horizontal rotor forces and moments for a pitch moment disturbance of 1 Nmm with a hinge offset of 10 mm.

flapping hinge offset leads to a faster buildup of larger rotor moments as compared to a rotor without offset.

Flapping due to forward velocity

To derive the flapping angles due to forward velocity of the helicopter, the same procedure is followed as for roll and pitch velocities: the moment equilibria around the flapping hinges of both rotors are formulated, sine and cosine terms are separated and the resulting two equations are solved for the respective flapping angles. However, in forward flight, only two moments need to be taken into account. These are again the centrifugal moment M^{cf} and the aerodynamic moment M^{aero}, the latter, however, in a different form because the helicopter is now in forward flight. Since the helicopter does not rotate in this case, no gyroscopic moment is present.

Starting with the upper rotor, the two moments are formulated as

$$M^{\text{cf}}_{\text{up}} = \Omega^2_{\text{up}} \frac{e_{\text{com}} M_{\text{bld}}}{g} \left(b^{\text{fwd}}_{1_{\text{up}}} \sin \psi_{\text{up}} + a^{\text{fwd}}_{1_{\text{up}}} \cos \psi_{\text{up}} \right), \quad (5.27)$$

$$\begin{aligned}
M^{\text{aero}}_{\text{up}} =& \frac{\rho}{2ac} (\Omega_{\text{up}} R)^2 \left(1 - \frac{e_{\text{com}}}{R}\right)^2 R^2 \Bigg\{ \\
& \left[2\theta_0 \mu \left(\frac{1}{3} + \frac{e_{\text{com}}}{6R} \right) + 2\theta_1 \mu \left(\frac{1}{4} - \frac{e_{\text{com}}}{6R} - \frac{1}{12} \left(\frac{e_{\text{com}}}{R} \right)^2 \right) \right. \\
& \left. + \frac{\mu}{2} \left(\mu \alpha_s - \frac{v_1}{\Omega_{\text{up}} R} \right) - a^{\text{fwd}}_{1_{\text{up}}} \left(\frac{1}{4} - \frac{\mu^2}{8} - \frac{e_{\text{com}}}{6R} - \frac{1}{12} \left(\frac{e_{\text{com}}}{R} \right)^2 \right) \right] \sin \psi_{\text{up}} \\
& + \left[b^{\text{fwd}}_{1_{\text{up}}} \left(\frac{1}{4} + \frac{\mu^2}{8} - \frac{e_{\text{com}}}{6R} - \frac{1}{12} \left(\frac{e_{\text{com}}}{R} \right)^2 \right) \right. \\
& \left. - \frac{v_1}{\Omega_{\text{up}} R} \left(\frac{1}{4} - \frac{e_{\text{com}}}{6R} - \frac{1}{12} \left(\frac{e_{\text{com}}}{R} \right)^2 \right) - \mu a_0 \left(\frac{1}{3} + \frac{e_{\text{com}}}{6R} \right) \right] \cos \psi_{\text{up}} \Bigg\}. \quad (5.28)
\end{aligned}$$

These moments have to equilibrate around the flapping hinge:

$$M^{\text{cf}}_{\text{up}} + M^{\text{aero}}_{\text{up}} = 0. \quad (5.29)$$

Separating the sine and cosine terms and solving simultaneously for the longitudinal and lateral flapping angles $a_{1\text{up}}^{\text{fwd}}$ and $b_{1\text{up}}^{\text{fwd}}$ leads to expressions of the form:

$$a_{1\text{up}}^{\text{fwd}} = \frac{\frac{8}{3}\theta_0 + 2\theta_1\mu + 2\mu\left(\mu\alpha_s - \frac{k^{\text{lift}}}{2\mu}\right)}{1 - \frac{\mu^2}{2}} + \frac{12\left(\frac{e}{R}\right)}{\gamma\left(1 - \frac{e}{R}\right)^3\left(1 - \frac{\mu^4}{4}\right)}$$
$$\left[k^{\text{lift}}/\sigma\left(\frac{\frac{8}{9}\frac{\mu\gamma}{a}\left(1-\frac{e}{R}\right)^2}{1+\frac{e}{2R}} + \frac{\sigma}{2\mu}\right)\right], \qquad (5.30)$$

$$b_{1\text{up}}^{\text{fwd}} = \frac{k^{\text{lift}}/\sigma}{1+\frac{\mu^2}{2}}\left(\frac{\frac{8}{9}\frac{\mu\gamma}{a}\left(1-\frac{e}{R}\right)^2}{1+\frac{e}{2R}} + \frac{\sigma}{2\mu}\right) + \frac{12\left(\frac{e}{R}\right)}{\gamma\left(1-\frac{e}{R}\right)^3\left(1-\frac{\mu^4}{4}\right)}$$
$$\left[\frac{8}{3}\theta_0\mu + 2\theta_1\mu + 2\mu\left(\mu\alpha_s - \frac{k^{\text{lift}}}{2\mu}\right)\right]. \qquad (5.31)$$

For the lower rotor, the same procedure is applied. Again, the two moments can be formulated according to the sign convention in Figure 3.4, which accounts for several sign changes in the otherwise identical expressions.

$$M_{\text{lo}}^{\text{c.f.}} = \Omega_{\text{lo}}^2 e_{\text{com}} \frac{M_b}{g}\left(b_{1\text{lo}}^{\text{fwd}}\sin\psi_{\text{lo}} - a_{1\text{lo}}^{\text{fwd}}\cos\psi_{\text{lo}}\right), \qquad (5.32)$$

$$M_{\text{lo}}^{\text{aero}} = \frac{\rho}{2ac}(\Omega_{\text{lo}}R)^2\left(1 - \frac{e_{\text{com}}}{R}\right)^2 R^2\Bigg\{$$
$$\left[2\theta_0\mu\left(\frac{1}{3} + \frac{e_{\text{com}}}{6R}\right) + 2\theta_1\mu\left(\frac{1}{4} - \frac{e_{\text{com}}}{6R} - \frac{1}{12}\left(\frac{e_{\text{com}}}{R}\right)^2\right)\right.$$
$$\left. + \frac{\mu}{2}\left(\mu\alpha_s - \frac{v_1}{\Omega_{\text{lo}}R}\right) + a_{1\text{lo}}^{\text{fwd}}\left(\frac{1}{4} - \frac{\mu^2}{8} - \frac{e_{\text{com}}}{6R} - \frac{1}{12}\left(\frac{e_{\text{com}}}{R}\right)^2\right)\right]\sin\psi_{\text{lo}}$$
$$+ \left[b_{1\text{lo}}^{\text{fwd}}\left(\frac{1}{4} + \frac{\mu^2}{8} - \frac{e_{\text{com}}}{6R} - \frac{1}{12}\left(\frac{e_{\text{com}}}{R}\right)^2\right)\right.$$
$$\left. - \frac{v_1}{\Omega_{\text{lo}}R}\left(\frac{1}{4} - \frac{e_{\text{com}}}{6R} - \frac{1}{12}\left(\frac{e_{\text{com}}}{R}\right)^2\right) - \mu a_0\left(\frac{1}{3} + \frac{e_{\text{com}}}{6R}\right)\right]\cos\psi_{\text{lo}}\Bigg\}, \qquad (5.33)$$

The moment equilibrium about the flapping hinge also holds for the lower rotor:

$$M_{\text{lo}}^{\text{cf}} + M_{\text{lo}}^{\text{aero}} = 0. \qquad (5.34)$$

The solution for the lower rotor longitudinal and lateral flapping angles $a_{1\text{lo}}^{\text{fwd}}$ and $b_{1\text{lo}}^{\text{fwd}}$ has the same structure as in Equations (5.30)–(5.31), but opposite signs for the cross coupling terms.

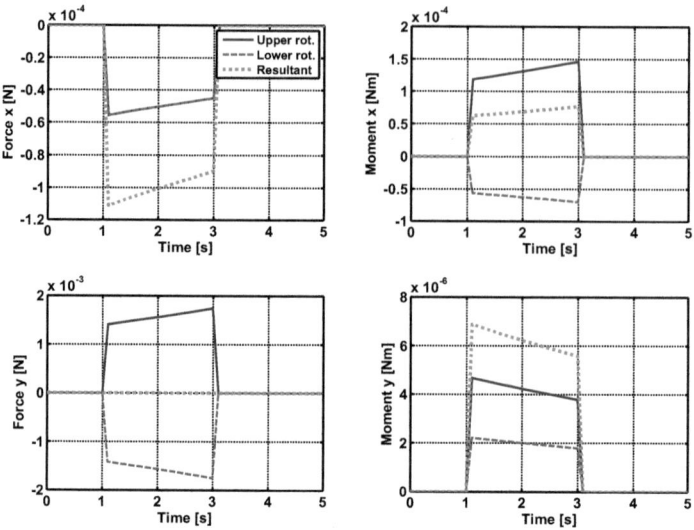

Figure 5.10: Horizontal rotor forces and moments for an x-force disturbance of 10 mN without hinge offset.

$$a_{1_{\text{lo}}}^{\text{fwd}} = \frac{\frac{8}{3}\theta_0 + 2\theta_1\mu + 2\mu\left(\mu\alpha_s - \frac{k^{\text{lift}}}{2\mu}\right)}{1 - \frac{\mu^2}{2}} - \frac{12\left(\frac{e}{R}\right)}{\gamma\left(1 - \frac{e}{R}\right)^3\left(1 - \frac{\mu^4}{4}\right)}$$

$$\left[k^{\text{lift}}/\sigma\left(\frac{\frac{8}{9}\frac{\mu\gamma}{a}\left(1 - \frac{e}{R}\right)^2}{1 + \frac{e}{2R}} + \frac{\sigma}{2\mu}\right)\right], \quad (5.35)$$

$$b_{1_{\text{lo}}}^{\text{fwd}} = \frac{k^{\text{lift}}/\sigma}{1 + \frac{\mu^2}{2}}\left(\frac{\frac{8}{9}\frac{\mu\gamma}{a}\left(1 - \frac{e}{R}\right)^2}{1 + \frac{e}{2R}} + \frac{\sigma}{2\mu}\right) - \frac{12\left(\frac{e}{R}\right)}{\gamma\left(1 - \frac{e}{R}\right)^3\left(1 - \frac{\mu^4}{4}\right)}$$

$$\left[\frac{8}{3}\theta_0\mu + 2\theta_1\mu + 2\mu\left(\mu\alpha_s - \frac{k^{\text{lift}}}{2\mu}\right)\right]. \quad (5.36)$$

Figure 5.10 shows the rotor forces and moments for an x-force disturbance of 10 mN of a rotor without hinge offset. While the x-forces counteract the disturbance with amplitudes in the same direction, the y-forces and respective x-moments react in opposite directions, canceling each other out. The buildup of rotor moments is not helpful to counteract a force disturbance, but is an inevitable result of blade flapping and the buildup of rotor forces. Hence, the force disturbance is not only rejected by a rotor force response, but also leads to a significant pitch moment. In Figure 5.11, the same result is shown for a flapping hinge offset of 10 mm. Qualitatively, the same result can be observed. While the cross-direction contributions cancel out, the force contributions on the disturbed axis counteract

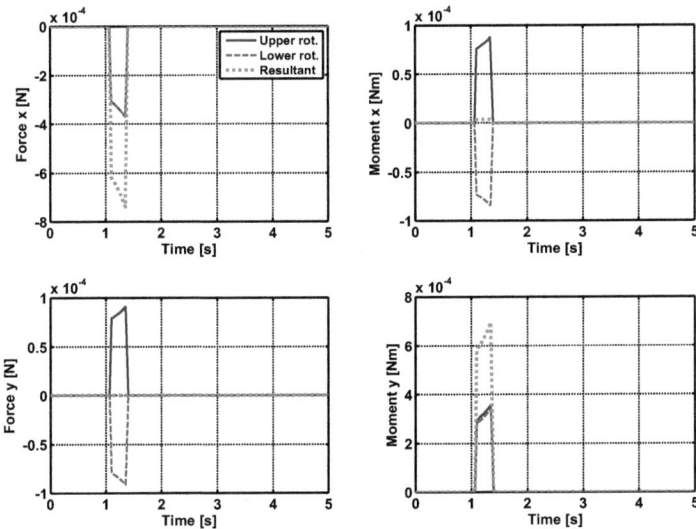

Figure 5.11: Horizontal rotor forces and moments for an x-force disturbance of 10 mN with a hinge offset of 10 mm.

the disturbance, and a pitch moment is generated. Since the hinge offset does not influence the magnitude of the rotor forces, a larger hinge offset does not lead to a stronger linear disturbance rejection. Figures 5.12–5.13 show the same results for a y-force disturbance. There, the restoring force and moment resultants are F_y and M_x, while F_x and M_y cancel out almost completely. Compared to a conventional single main rotor, the coaxial rotor results in stronger force disturbance rejection (two rotors yield a restoring force instead of one), while the undesired cross terms are canceled out as a result of the opposite rotation directions of the rotors. However, the generation of the restoring force inevitably also creates a moment, which is undesired for force disturbance rejection. Because of this, also the introduction of a hinge offset is not advantageous for force disturbance rejection, since it introduces larger rotor moments.

5.3 Stabilizer bar module

The simulation model includes the possibility to simulate a stabilizer bar linked to the upper rotor. The functionality of this device is investigated in detail in Chapter 4. Based on the results obtained from that analysis, the stabilizer bar modeling for the actual helicopter simulation can be kept rather simple. The stabilizer bar following time T_f, which has been derived from multi-body simulation and experiments in Chapter 4, is used for the module.

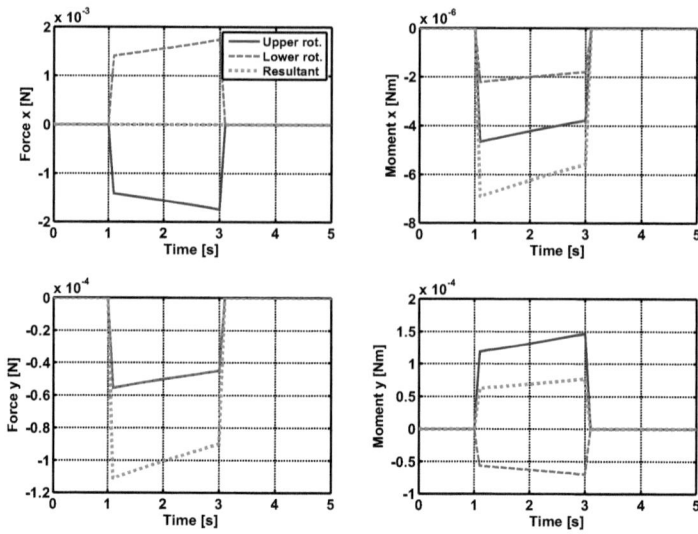

Figure 5.12: Horizontal rotor forces and moments for a y-force disturbance of 10 mN without hinge offset.

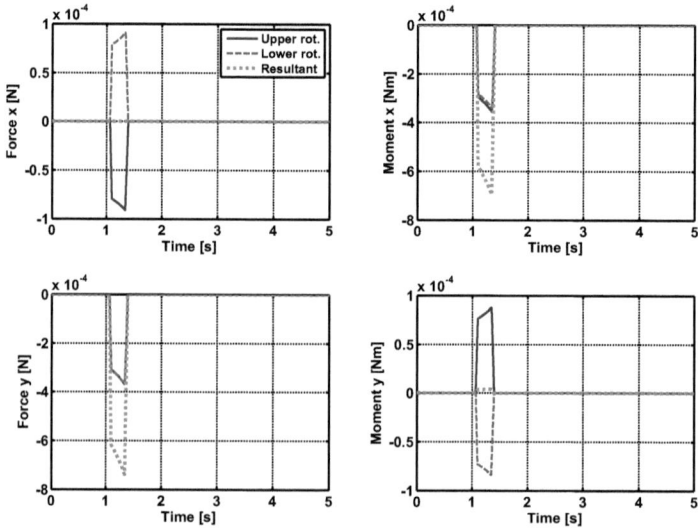

Figure 5.13: Horizontal rotor forces and moments for a y-force disturbance of 10 mN with a hinge offset of 10 mm.

In general, the stabilizer bar can be treated as a first order element. It is important to observe, that the stabilizer bar does not react to perturbations in the roll and pitch angles of the helicopter, but to their rates [63]. Otherwise, if subjected to a constant roll or pitch angle displacement, the flapping motion of the stabilizer bar would never end, which is obviously not the case. Hence, under the assumption of a correctly tuned phase angle with respect to the pitch axis of the blades, the flapping angle of the stabilizer bar can be described as a harmonic function with the longitudinal and lateral stabilizer bar flapping angles η and ξ as coefficients:

$$\beta_{\text{sb}} = -\eta \cos \psi_{\text{up}} - \xi \sin \psi_{\text{up}}. \tag{5.37}$$

Due to this harmonic function, the roll and pitch axis of the helicopter can be treated separately. The magnitudes of the coefficients η and ξ are described by the following first order differential equation, which has the body-fixed angular rates p and q as inputs:

$$\begin{bmatrix} \dot{\eta} \\ \dot{\xi} \end{bmatrix} = -\frac{1}{T_{\text{f}}} \begin{bmatrix} \eta \\ \xi \end{bmatrix} - \begin{bmatrix} p \\ q \end{bmatrix}, \quad \begin{bmatrix} \eta \\ \xi \end{bmatrix} = \kappa \cdot \begin{bmatrix} A_{1_{\text{up}}} \\ B_{1_{\text{up}}} \end{bmatrix}. \tag{5.38}$$

The coefficients yielded by this equation are equal to the lateral and longitudinal steering inputs of the upper rotor, scaled by the gearing ratio κ that has been introduced in Chapter 4. It represents the linearized push rod kinematics between the flapping degree of freedom of the stabilizer bar and the pitching degree of freedom of the blades.
Simulation results for the stabilizer bar module are given in Section 5.6.

5.4 Swash plate steering module

If the helicopter is steered by a swash plate, this leads to a cyclic pitch input on the rotor blades, i.e. the individual blade pitch depends on the rotor azimuth position of the blade. This varying blade pitch introduces two additional moments to the previously introduced equilibrium in Equation (5.24) for the lower rotor. These are the swash plate aerodynamic moments in the longitudinal and lateral direction $M^{\text{aero,sp,long}}$ and $M^{\text{aero,sp,lat}}$.

$$M^{\text{aero,sp,long}} = \frac{\rho}{2ac} (\Omega_{\text{lo}} R)^2 \left(1 - \frac{e_{\text{com}}}{R}\right)^2 R^2$$
$$\left\{ -B_{1_{\text{lo}}} \left(\frac{1}{4} + \frac{3}{8}\mu^2 + \frac{e_{\text{com}}}{6R} + \frac{1}{12}\left(\frac{e_{\text{com}}}{R}\right)^2 \right) \right\} \sin \psi_{\text{lo}}, \tag{5.39}$$

$$M^{\text{aero,sp,lat}} = \frac{\rho}{2ac} (\Omega_{\text{lo}} R)^2 \left(1 - \frac{e_{\text{com}}}{R}\right)^2 R^2$$
$$\left\{ A_{1_{\text{lo}}} \left(\frac{1}{4} + \frac{\mu^2}{8} + \frac{e_{\text{com}}}{6R} + \frac{1}{12}\left(\frac{e_{\text{com}}}{R}\right)^2 \right) \right\} \cos \psi_{\text{lo}}. \tag{5.40}$$

They contain the control inputs $A_{1_{\text{lo}}}$ and $B_{1_{\text{lo}}}$, which are the longitudinal and lateral swash plate tilting angles. The moment equilibrium about the flapping hinge is formulated as

$$M^{\text{cf}} + M^{\text{aero}} + M^{\text{aero,sp,long}} + M^{\text{aero,sp,lat}} = 0, \tag{5.41}$$

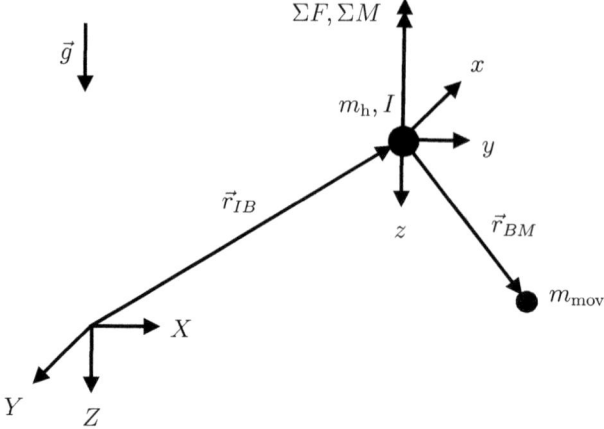

Figure 5.14: Schematic view of the center of mass displacement steering mechanism for Euler-Lagrange modeling.

where M^{cf} and M^{aero} are taken from Equations (5.32)–(5.33), respectively. Separation of the sine and cosine terms, simultaneous solution for the longitudinal and lateral flapping angles and subtraction of the terms that already contribute to flapping due to forward velocity, lead to the actively excited flapping contributions:

$$a_{1_{\text{lo}}}^{\text{sp}} = \frac{B_{1_{\text{lo}}}\left(1 + \frac{3}{2}\mu^3\right)}{1 - \frac{\mu^2}{2}}, \qquad a_{1_{\text{up}}}^{\text{sb}} = \frac{B_{1_{\text{up}}}\left(1 + \frac{3}{2}\mu^3\right)}{1 - \frac{\mu^2}{2}}, \qquad (5.42)$$

$$b_{1_{\text{lo}}}^{\text{sp}} = \frac{A_{1_{\text{lo}}}\left(1 + \frac{3}{2}\mu^3\right)}{1 - \frac{\mu^2}{2}}, \qquad b_{1_{\text{up}}}^{\text{sb}} = \frac{A_{1_{\text{up}}}\left(1 + \frac{3}{2}\mu^3\right)}{1 - \frac{\mu^2}{2}}. \qquad (5.43)$$

With this, the governing equations for swash plate steering, which are implemented in the module, are complete.

5.5 Center of mass displacement steering module

To displace the center of mass of a helicopter, a significant fraction of its total mass has to be moveable with respect to the helicopter fuselage. Typically, this mass is the battery of the helicopter, as on the μFR [7] or on the CoaX 2 [16, 25]. Since center of mass displacement steering is essentially a two body problem, again the Euler-Lagrange approach introduced in Chapter 4 is used for modeling. The modeling essentially follows the approach developed in [60].

A schematic view of the problem is shown in Figure 5.14. The position of the helicopter fixed coordinate fame in the inertial frame is described by the vector $\vec{r}_{IB}$, the position of the moveable mass m_{mov} with respect to the helicopter fixed frame is described by the vector $\vec{r}_{BM}$. The gravity vector $\vec{g}$ is directed parallel to the inertial Z-axis. All forces and moments on the helicopter, including rotor forces and moments, are introduced as external forces denoted ΣF and ΣM.

Base point of the modeling is the Lagrange equation (4.3), which is repeated here:

$$\frac{\mathrm{d}}{\mathrm{d}t}\left(\frac{\partial L}{\partial \dot{\vec{q}}}\right)^{\mathrm{T}} - \frac{\partial L}{\partial \vec{q}}^{\mathrm{T}} - \vec{F}_{\text{gen}} = 0.$$

The set of generalized coordinates $\vec{q}$ that fully describes the two body system is chosen as the position of the helicopter in the inertial coordinate frame, and the roll, pitch and yaw angles:

$$\vec{q} = [X, Y, Z, \phi, \theta, \psi]^{\mathrm{T}}. \tag{5.44}$$

With this, first of all the translational kinetic energy of the system is formulated. It consists of the contributions of the helicopter and the moveable mass:

$$T^{\text{trans}} = T_{\text{h}}^{\text{trans}} + T_{\text{mov}}^{\text{trans}}$$

$$= \frac{1}{2} m_{\text{h}} \cdot \left|\frac{\mathrm{d}}{\mathrm{d}t}(\vec{r}_{\text{IB}})\right|^2 + \frac{1}{2} m_{\text{mov}} \cdot \left|\frac{\mathrm{d}}{\mathrm{d}t}(\vec{r}_{\text{IB}} + \mathbf{A}_{\text{IB}} \cdot \vec{r}_{\text{BM}})\right|^2. \tag{5.45}$$

Here, the time derivatives of the vectors $\vec{r}_{\text{IB}}$ and $\vec{r}_{\text{BM}}$ denote the translational velocities of the helicopter and the moveable mass, while the matrix $\mathbf{A}_{IB}$ is a rotational transformation matrix from the helicopter fixed to the inertial frame in agreement with Equation (3.2).

For the rotational kinetic energy, helicopter and moveable mass are treated as one body with cumulated inertia tensor to reduce model complexity. This simplification can be made, since the rotation of the two bodies with respect to each other is very small and does not significantly contribute to the total kinetic energy. Furthermore assuming a diagonal inertia tensor of the form

$$\mathbf{I} = diag\left([I_{xx}, I_{yy}, I_{zz}]^{\mathrm{T}}\right), \tag{5.46}$$

the rotational kinetic energy of the system accounts for

$$T^{\text{rot}} = \frac{1}{2} \cdot \left(\mathbf{R}_{BI} \begin{bmatrix} \dot{\phi} \\ \dot{\theta} \\ \dot{\psi} \end{bmatrix}\right)^{\mathrm{T}} \cdot \mathbf{I} \cdot \mathbf{R}_{BI} \begin{bmatrix} \dot{\phi} \\ \dot{\theta} \\ \dot{\psi} \end{bmatrix}. \tag{5.47}$$

Then, the total kinetic energy is the sum of translational and the rotational energy:

$$T = T^{\text{trans}} + T^{\text{rot}}. \tag{5.48}$$

For the potential energy, the potentials of the helicopter mass m_{h} and the moveable mass m_{mov} are described in the gravitational field with respect to the inertial coordinate frame. These potentials are simply

$$V = V_{\text{h}} + V_{\text{mov}}$$

$$= -m_{\text{h}} \cdot \vec{r}_{\text{IB}}^{\mathrm{T}} \cdot \vec{g} - m_{\text{mov}} \left(\vec{r}_{\text{IB}} + \mathbf{A}_{\text{IB}} \cdot \vec{r}_{\text{BM}}\right)^{\mathrm{T}} \cdot \vec{g}. \tag{5.49}$$

In order to introduce the external forces and moments, they have to be generalized by multiplication with the Jacobian matrices (see Equation (4.41)). For the system at hand, the Jacobians conveniently yield

$$\mathbf{J}^f = \frac{\partial}{\partial \vec{q}}\left(\frac{\mathrm{d}}{\mathrm{d}t}(\vec{r}_{\mathrm{IB}})\right) = \begin{bmatrix} 1 & 0 & 0 & 0 & 0 & 0 \\ 0 & 1 & 0 & 0 & 0 & 0 \\ 0 & 0 & 1 & 0 & 0 & 0 \end{bmatrix}, \quad (5.50)$$

$$\mathbf{J}^t = \frac{\partial}{\partial \vec{q}}\left(\frac{\mathrm{d}}{\mathrm{d}t}\left(\begin{bmatrix} \phi & \theta & \psi \end{bmatrix}^{\mathrm{T}}\right)\right) = \begin{bmatrix} 0 & 0 & 0 & 1 & 0 & 0 \\ 0 & 0 & 0 & 0 & 1 & 0 \\ 0 & 0 & 0 & 0 & 0 & 1 \end{bmatrix}. \quad (5.51)$$

Hence, the Lagrange equation (4.3) is complete and can be solved obtain the six modified equations of motion for center of mass displacement steering of the helicopter. The expressions for these equations are too lengthy to be displayed here in a meaningful manner. Principally, they are functions of the helicopter mass m_{h}, the moveable mass m_{mov}, the accumulated inertia tensor of the helicopter $\mathbf{I}$ and the trajectory of the moveable mass with respect to the helicopter characterized by the vector $\vec{r}_{\mathrm{BM}}$ and its first and second time derivatives. Very simple input trajectories for the moveable mass have been chosen for simulation. For a more sophisticated input trajectory design, approaches are given in [45, 46].

The equations of motion are significantly complicated by including the translational kinetic energy of the moveable mass $T_{\mathrm{mov}}^{\mathrm{trans}}$ into the Lagrangian approach. Leaving out this term leads to simpler, yet still undemonstrable equations of motion. Figure 5.15 shows a comparison of both sets of equations of motion for the same input trajectory. The error due to the simplification is relatively small for simulation times up to 2 s. Zooming into the y-trajectory, as it is done in Figure 5.16, reveals the largest difference between the two sets of equations of motion. The equations incorporating the linear kinetic energy of the moveable mass show a slight inverse response behavior as a result of linear momentum exchange between moveable mass and helicopter: an acceleration of the moveable mass to the right accelerates the helicopter fuselage to the left, until the rolling motion sets in and the complete system starts moving to the right. It has to be noted that this effect is very weak and certainly not very relevant for the complete time trajectory of the helicopter. On the other hand, the calculation time for the simpler model is strongly reduced. Hence, it makes sense to use the simpler equations of motion, which is done in the simulation results section and in the quantitative comparison between the two steering principles. These modified equations of motion constitute the equations of motion module for center of mass displacement steering according to Figure 5.14.

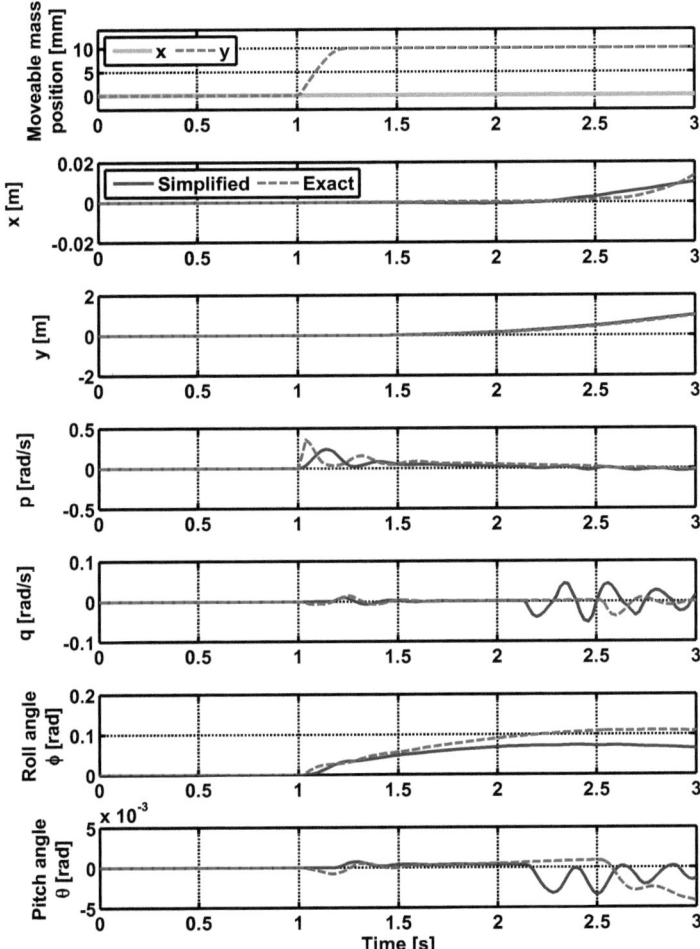

Figure 5.15: Horizontal linear and angular helicopter trajectories for a CM roll steering input of 10 mm to the exact and the simplified CM steering model.

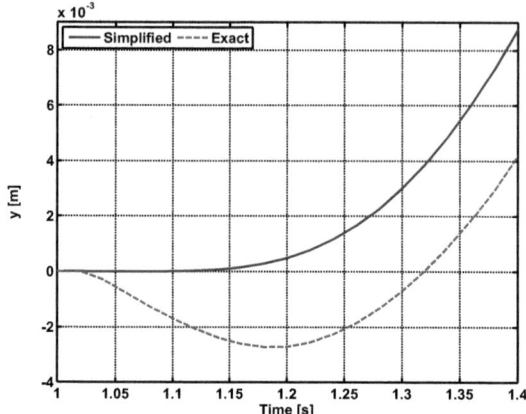

Figure 5.16: Zoom into the y-displacement to show the inverse response behavior of the exact model.

5.6 Simulation results

In this section, the simulation results for different possible setups of the coaxial helicopter are shown. First of all, a passive helicopter without steering inputs subjected to a roll moment disturbance is simulated without and with stabilizer bar. Secondly, swash plate steering is simulated without and with stabilizer bar, and without and with flapping hinge offset. Finally, center of mass displacement steering is simulated and compared to swash plate steering, with a stabilizer bar for both steering options.

For all simulations, heave and yaw dynamics have been disabled, since the major interest of this work lies in the horizontal responses of the system to evaluate steering and stabilization performance. Hence, all simulation results show horizontal linear and angular displacement trajectories and the body angular rates p and q.

5.6.1 Stabilizer bar

To show the influence of the stabilizer bar on the model, three different configurations are chosen for simulation: no stabilizer bar, a stabilizer bar with a following time of $T_f = 250\,\mathrm{ms}$ and a stabilizer bar with a following time of $T_f = 500\,\mathrm{ms}$, all configurations without active steering inputs and subjected to a rectangular roll moment disturbance of $30\,\mathrm{Nmm}$ at a simulation time of $1\,\mathrm{s}$. Figure 5.17 shows the time trajectories for all three configurations. While the configuration without stabilizer bar becomes unstable and drifts away at high roll rate, both configurations with stabilizer bar stabilize after the disturbance. For the latter, two observations are important: due to the hinge offset of $5\,\mathrm{mm}$, which introduces cross

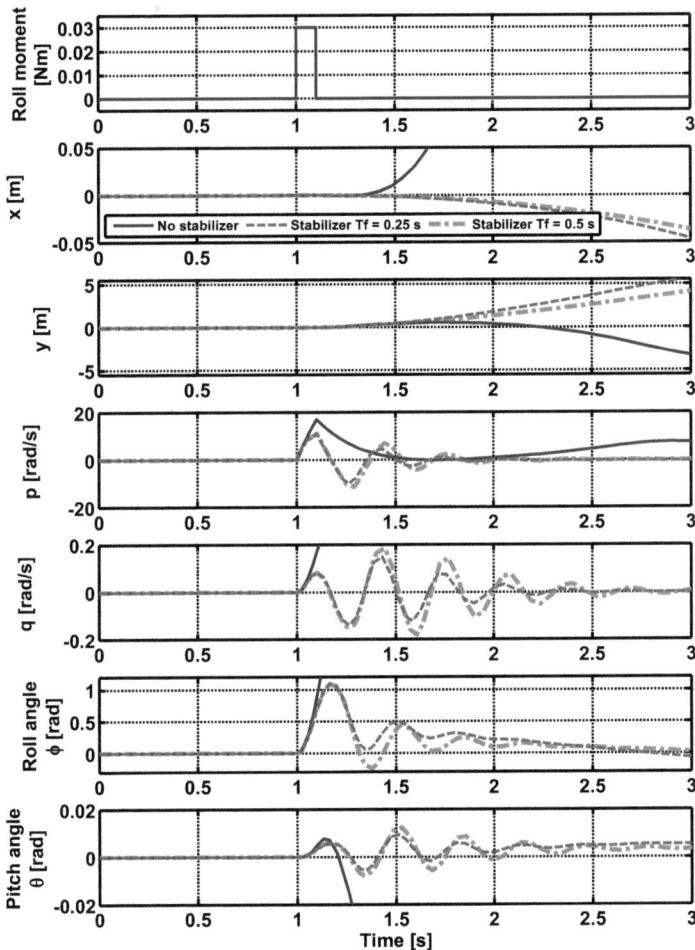

Figure 5.17: Horizontal linear and angular helicopter trajectories for a roll moment disturbance of 30 Nmm with a hinge offset of 5 mm and different stabilizer bar parametrizations.

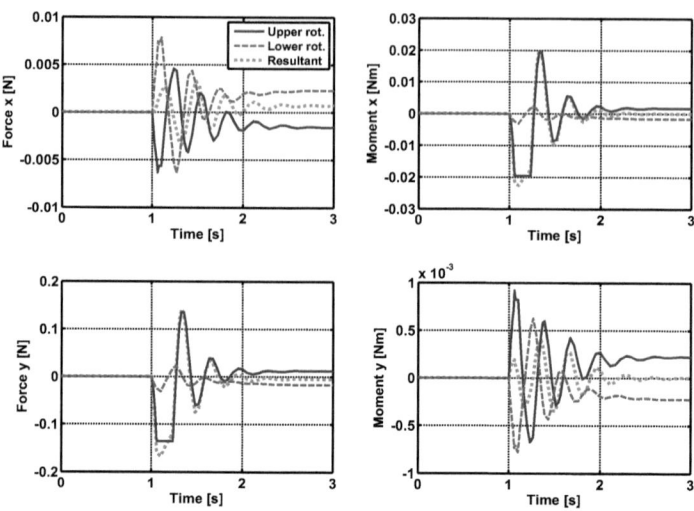

Figure 5.18: Horizontal rotor forces and moments for a roll moment disturbance of 30 Nmm with a hinge offset of 5 mm and a stabilizer bar with following time $T_f = 250$ ms.

coupling terms on the rotors, the disturbance also affects the initially undisturbed pitch angle. Secondly, a larger following time does not necessarily lead to better stabilization. For the following time of 500 ms, which is a slower stabilizer bar, the angular amplitudes are damped worse than for the faster following time of 250 ms. Basically, if the stabilizer is too slow to follow the oscillation frequency of the helicopter fuselage, it can even destabilize the helicopter due to out-of-phase behavior.

For further investigation of this observation, the rotor forces and moments are shown in Figure 5.18 for the low and in Figure 5.19 for the high stabilizer bar following time. Of special interest are the moment trajectories. Due to the action of the stabilizer bar, the x-moment is clearly dominated by the upper rotor, and both rotors initially act in phase. For the y-moment, initially both rotors act out of phase, which minimizes cross coupling time. However, after approximately half a period of oscillation, the rotors begin shifting their phase towards an in-phase behavior, which means that cross coupling cancelation is shifted towards cross coupling amplification. This effect is worse for the higher following time.

In summary, these results show that the stabilizer bar can be successfully used as a passive stabilization device. However, if the stabilizer following time is chosen too small or too large, the system will destabilize despite the use of a stabilizer bar. In the former case this will happen due to a lack of action, in the latter due to phase shifting between the flapping motions of the two rotors.

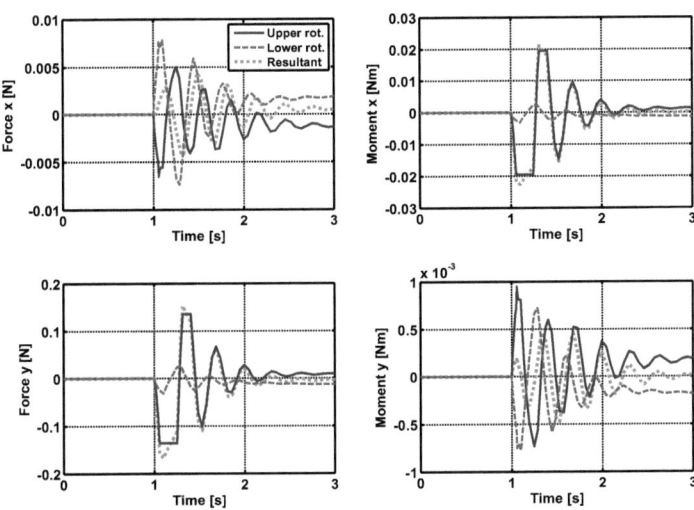

Figure 5.19: Horizontal rotor forces and moments for a roll moment disturbance of 30 Nmm with a hinge offset of 5 mm and a stabilizer bar with following time $T_f = 500$ ms.

5.6.2 Swash plate steering

To study swash plate steering, the simulation is run for the helicopter without and with a stabilizer bar at following time $T_f = 250$ ms, and without and with a hinge offset of 10 mm. The steering input is a positive roll of 5° at $t = 1$ s, which is maintained until the end of the simulation.

Figure 5.20 shows the helicopter trajectories and the steering input for the helicopter without stabilizer bar, without hinge offset and with a hinge offset of 10 mm. As it can be expected, the reaction of the helicopter to the steering input is mostly on the roll axis and in the y-displacement. The reaction becomes a bit more aggressive with a larger hinge offset, since the steering moments become larger. Also, the larger hinge offset induces more cross coupling. Inspection of the respective rotor forces and moments in Figure 5.21 for no hinge offset and in Figure 5.22 for 10 mm hinge offset shows that the passive flapping dynamics of the rotors counteract the steering input. The upper rotor acts completely in the opposite direction of the desired steering input, since only the passive flapping dynamics contribute. On the lower rotor, which is actively steered by the swash plate, the contributions of the active input and the passive flapping dynamics compete, with their resultant still pointing in the desired steering direction. The introduction of the hinge offset does not significantly change the resultant forces and moments, since both acting and counteracting effects are amplified. Finally, it must be noted that the simulation has been stopped after 3 s due to destabilization.

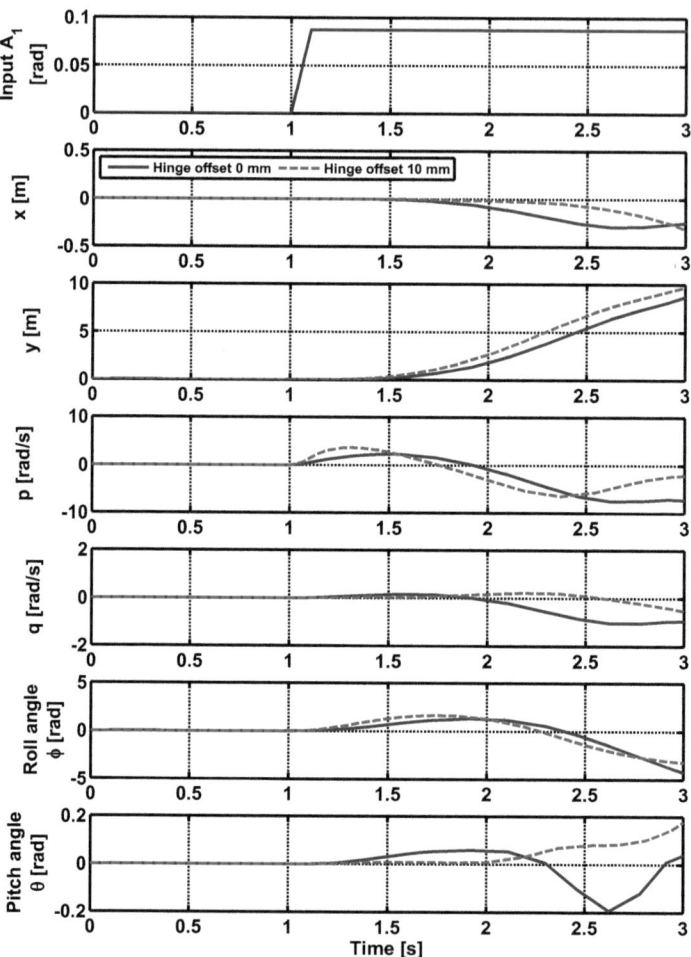

Figure 5.20: Horizontal linear and angular helicopter trajectories for a roll steering input of 5° with different hinge offsets and without stabilizer bar.

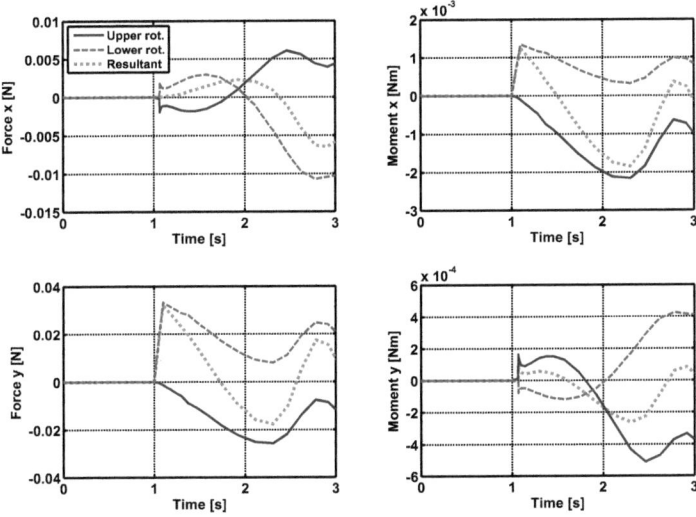

Figure 5.21: Horizontal rotor forces and moments for a roll steering input of 5° without hinge offset and without stabilizer bar.

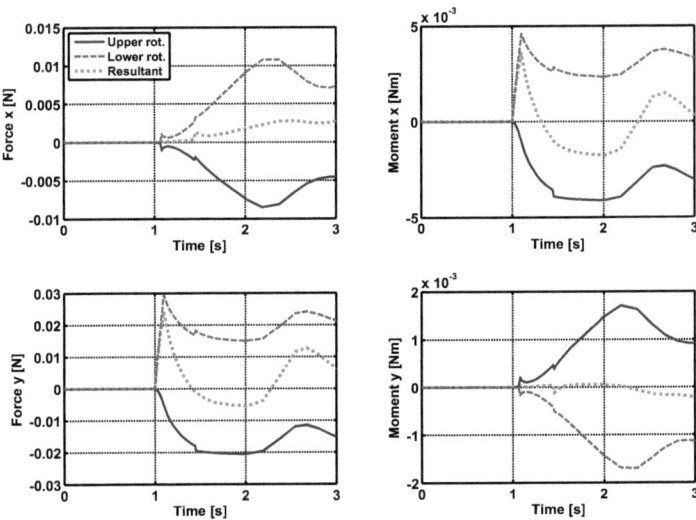

Figure 5.22: Horizontal rotor forces and moments for a roll steering input of 5° with a hinge offset of 10 mm and without stabilizer bar.

Figure 5.23 shows the helicopter trajectories and the steering input for the helicopter with a stabilizer bar at $T_f = 250$ ms, without hinge offset and with a hinge offset of 10 mm. The most important difference to the trajectories without stabilizer bar in Figure 5.20 is that due to the action of the stabilizer bar, the steering is counteracted even stronger. This becomes most obvious in the y-displacement at $t_{sim} = 3$ s, which is smaller than for the model without stabilizer bar at the end of the simulation. The rotor forces and moments in Figure 5.24 and Figure 5.25 are in agreement with this. The counteracting effect of the upper rotor has become stronger, and also the oscillation of the stabilizer bar can be seen in the y-force and x-moment trajectories.

In summary, these results show that the previously mentioned stability advantage of the coaxial rotor configuration is a disadvantage when it comes to active steering of the helicopter. Maneuvering is possible, however, fast and agile maneuvers are suppressed by the stability of the rotor system. This becomes even worse, if a stabilizer bar is used.

5.6.3 Center of mass displacement steering

The simulation for center of mass displacement steering is executed with the simplified equations of motion according to Section 5.5. As for the swash plate steering simulations, the steering input is given on the roll axis to achieve positive rolling of the helicopter. The simulation is run for rotors without and with 10 mm hinge offset and with stabilizer bar. For center of mass displacement steering, the rotor forces and moments are not displayed, because their qualitative behavior corresponds to that for a roll disturbance moment: both rotors counteract the roll motion and the linear motion in the y-direction. Figure 5.26 shows the simulation result for a horizontal moveable mass displacement of 10 mm in the helicopter fixed y-direction. After application of the input signal at $t = 1$ s, the helicopter displaces in the y-direction and stabilizes at a roll angle of approximately 0.1 rad for the rotors without hinge offset, and at a roll angle of about 0.8 rad for the rotors with a hinge offset of 10 mm. Again, due to the partial cancelation of the cross coupling terms of both rotors, the motion amplitudes in the direction of the axis that is not actuated are rather small. The lateral displacement is with approximately 1.6 m at the end of the simulation time much smaller than what could be achieved with a swash plate at full stroke. Already at the moderate stroke of 5°, the swash plate reaches a lateral displacement between 20 m and 30 m, depending on the hinge offset.

5.6.4 Comparison

With the modular dynamic model it is possible to principally compare the steering performances of swash plate and center of mass displacement steering. In order to do so, center of mass displacement steering is simulated with the steering mechanism at full stroke for a roll maneuver. This means that the moveable mass is displaced by 10 mm. The swash

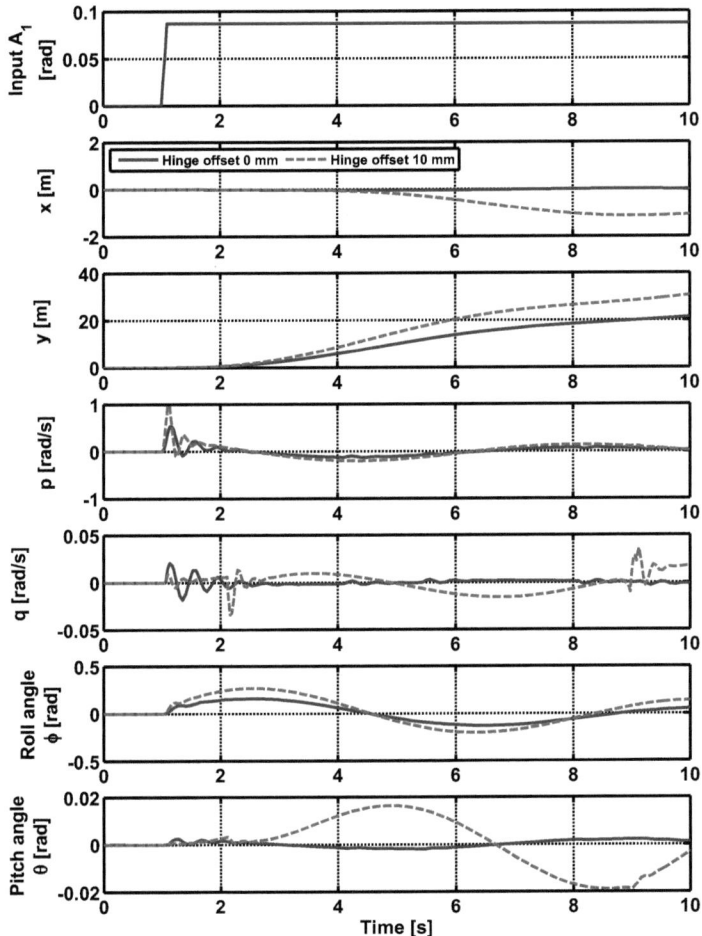

Figure 5.23: Horizontal linear and angular helicopter trajectories for a roll steering input of 5° with different hinge offsets and with stabilizer bar.

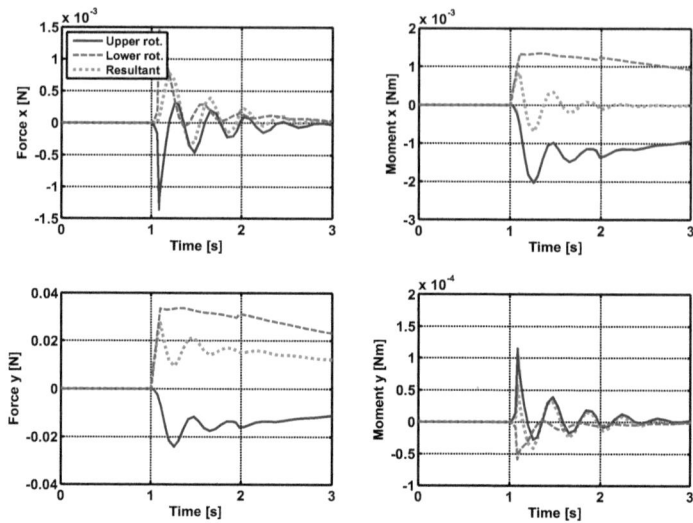

Figure 5.24: Horizontal rotor forces and moments for a roll steering input of 5° with a hinge offset of 0 mm and with stabilizer bar.

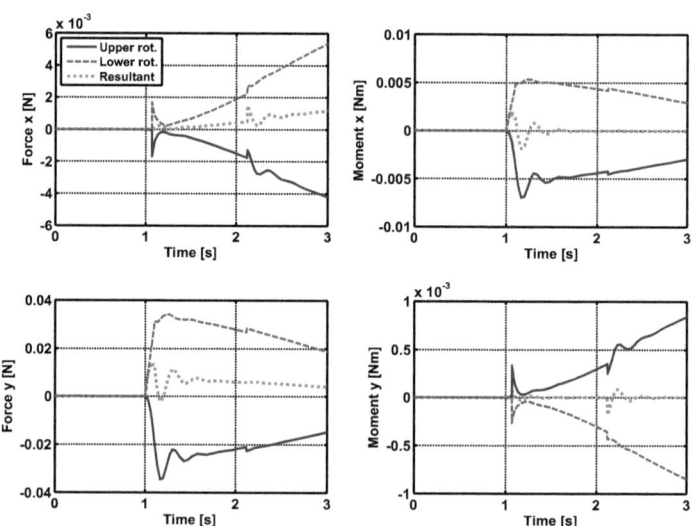

Figure 5.25: Horizontal rotor forces and moments for a roll steering input of 5° with a hinge offset of 10 mm and with stabilizer bar.

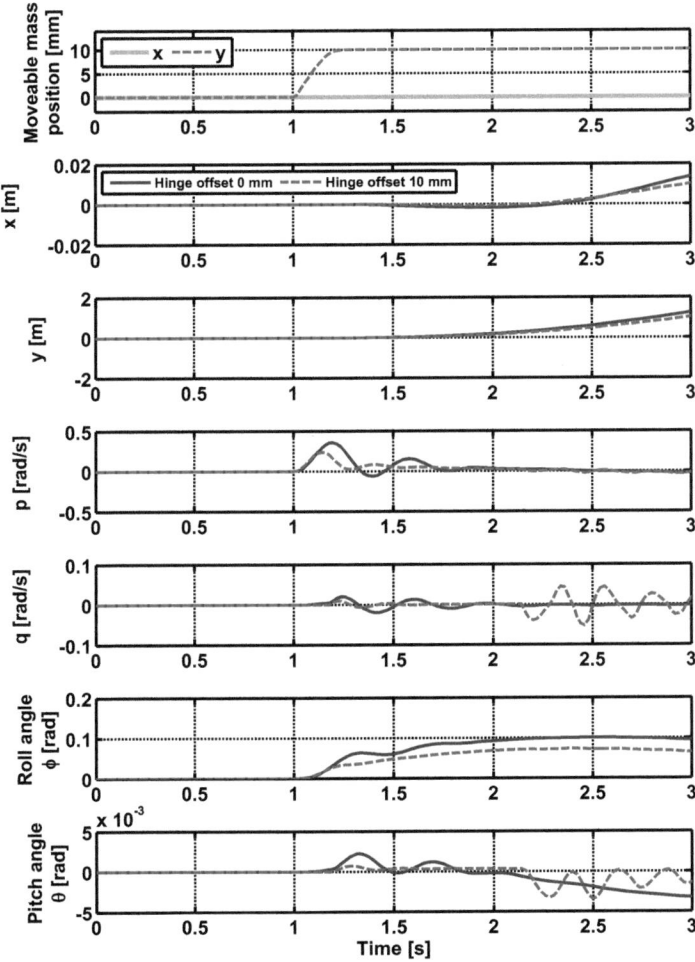

Figure 5.26: Horizontal linear and angular helicopter trajectories for a CM roll steering input of 10 mm to the simplified CM steering model with different hinge offsets.

Table 5.1: Parameters of the steering principle comparison.

Parameter	Value	Unit	Description
a_{max}	2.15	[m/s²]	Max. acceleration of mass
c	17.4 × 10^{-3}	[m]	Chord length
k^{drag}	2	[-]	Flat plate drag coefficient
d	5×10^{-3}	[m]	Dist. pitch axis servo pushrod
h	5×10^{-3}	[m]	Dist. pitch axis blade center
m_{mov}	0.0127	[kg]	Moveable mass (muFly battery)
R	87.5 × 10^{-3}	[m]	Rotor radius
ρ	1.2	[kg/m³]	Air density
θ_{cm}	55	[°]	CM servo stroke
θ_{sp}	5	[°]	SP servo stroke
Ω	400	[rad/s]	Rotor speed

plate steering mechanism is actuated at an angle, which leads to approximately the same output in terms of the lateral y-displacement. This angle is at 5° for roll steering, which is only half of the possible full swash plate stroke of 10°. The corresponding input and helicopter trajectories are shown in Figure 5.27. Although the swash plate is not at full stroke, these control inputs on the two mechanisms are chosen to make an approximate performance comparison between the two.

The comparison is made in terms of the energy required for one stroke to achieve the outputs according to Figure 5.27. Starting with center of mass displacement steering, and noting that the maximal acceleration of the moveable mass according to the simulation results accounts for $a_{max} = 2.15 \, \text{m/s}^2$, while for full stroke the servo motor needs to deliver an angular displacement of 55°, the required work normalized by the servo lever arm yields

$$W_{cm,norm} = a_{max} m_{mov} \theta_{cm} = 2.62 \times 10^{-2} \, \text{Nrad}.$$

For swash plate steering, the approximate torque that needs to be overcome by the servo is the aerodynamic torque of the blade in the air flow. It can be calculated as

$$T_{sp} = \frac{1}{4} k^{drag} \rho \Omega^2 R^3 h c \sin(\theta_{sp})$$

with the parametrization from Table 5.1. Introducing the pitch lever length d, the respective force becomes

$$F_{sp} = \frac{T_{sp}}{d},$$

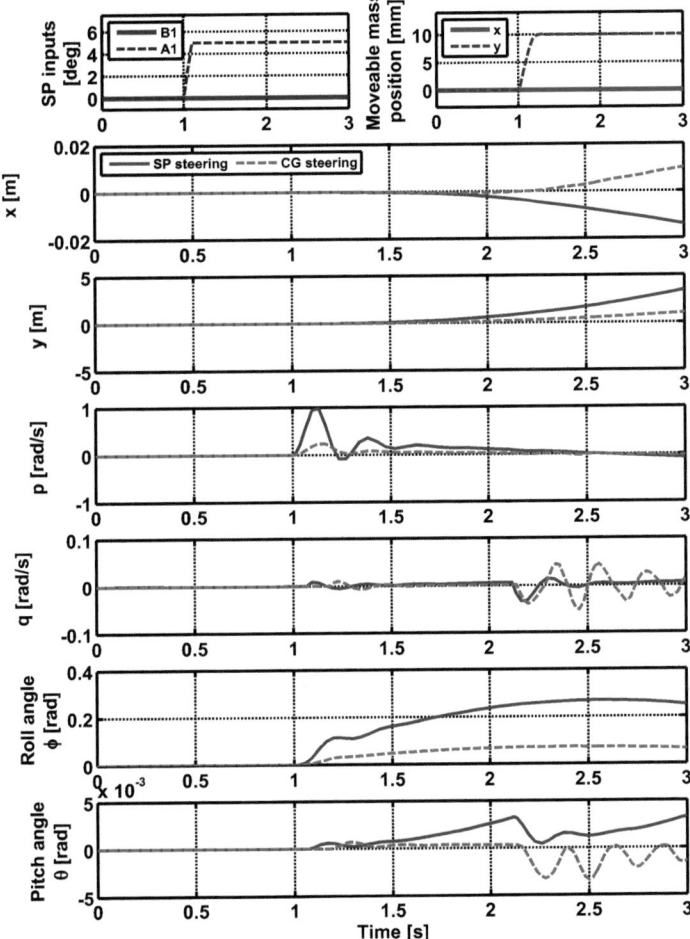

Figure 5.27: Horizontal linear and angular helicopter trajectories for swash plate and center of mass displacement steering.

and the normalized work for swash plate steering yields

$$W_{\text{sp,norm}} = F_{\text{sp}}\theta_{\text{sp}} = 8.51 \times 10^{-3}\,\text{Nrad}.$$

Hence, the ratio of the two normalized works accounts for

$$\frac{W_{\text{cm,norm}}}{W_{\text{sp,norm}}} \approx 3.$$

It shows that the equivalent of a center of mass shifting maneuver to a swash plate maneuver consumes approximately three times more energy. Taking further into account that for the inspected maneuver, center of mass displacement steering has no stroke reserve left, the superiority of the swash plate as a steering mechanism becomes obvious.

5.7 Summary

In this chapter, a modular dynamic simulation model for coaxial micro helicopters is presented. Modularity is achieved by separate modeling of the subsystems most relevant to the attitude stability of the helicopter, in order to allow for their activation and deactivation. These are the two rotors, the stabilizer bar, and two different steering principles, which are swash plate steering and steering by displacement of the center of mass of the helicopter. The complete model is implemented in Matlab/Simulink.

Core of the simulation model are the detailed rotor modules, which, in contrast to all currently used simulation models, also take into account the passively excited blade flapping of the rotors effected by roll, pitch and forward velocities of the helicopter. Moreover, these flapping dynamics are formulated in terms of direct helicopter design parameters, which makes the model valuable for dynamic characterization of the helicopter in the design phase.

The simulation results show that the coaxial rotor configuration offers a stability advantage over other rotor configurations, since disturbances are countered by both rotors, while their cross coupling terms almost cancel out due to the opposite directions of rotation. This advantage, however, comes at the price of less dynamic steering, since for the rotor system there is principally no difference between a disturbance and a steering input. Hence, the coaxial rotor configuration should be used in scenarios where its advantages can be exploited, for instance if long hovering time, only moderate flight maneuvers and compactness of the helicopter are required. Certainly, the coaxial configuration is not a candidate for helicopter aerobatics.

Further simulations show that the optional stabilizer bar is an effective device for passive stabilization of the helicopter. Nevertheless, exact tuning is necessary. Results for a very stable stabilizer, i.e. with a long following time, show that it can lead to a worse total stabilization of the helicopter.

Finally, in this chapter a qualitative and quantitative performance comparison between swash plate and center of mass displacement steering is made. The results state that the swash plate is clearly the better choice, in terms of steering bandwidth and amplitude as well as in terms of the energy required for the single steering action. The only case, for which center of mass steering can be envisioned, is if there is no other choice left. This could be the case if the helicopter is so small that a swash plate cannot be mechanically realized.

Chapter 6

Parameter study for passive roll and pitch stability

The previous chapter proposes a dynamic model for coaxial micro helicopters including a stabilizer bar on the upper rotor to passively stabilize the roll and pitch dynamics of the helicopter. The objective of this chapter is to investigate the possibility of achieving passive roll and pitch stability for a coaxial micro helicopter without a stabilizer bar. After all, the model structures in Chapter 5 show that the rotors with their passive flapping dynamics act as feedback loops on the helicopter equations of motion. By studying the influence of certain design parameters, which characterize these feedback loops, it may be possible to find parameter sets that lead to passive roll and pitch stabilization by exploitation of the passively excited blade flapping, without active control, a stabilizer bar or other passive stabilization devices [66]. Hence, these parameter sets could be used to build a rotor system that makes the stabilizer bar obsolete.

In the following Section 6.1, the model structure and the selection of the parameters to be varied are shown. Section 6.2 gives details on the software implementation of the parameter study, and the results of the different parameter variations are presented in Section 6.3.

6.1 Basic model and parameter selection

The model structure, on which the parameter study is based, is shown in Figure 6.1. It is derived from the model structures that have been introduced in Chapter 5. However, the stabilizer bar module has been removed as well as any steering input, such that the system at hand is a purely passive helicopter in terms of steering. Therefore, the roll and pitch stability of the helicopter can only be inherent, resulting from the passively excited flapping contributions of each rotor due to roll, pitch, and forward velocities. In order to disturb the model in hovering flight, the external forces and moments block is used to apply a roll or pitch disturbance moment.

Throughout the parameter study, a standard roll disturbance moment is used to disturb the

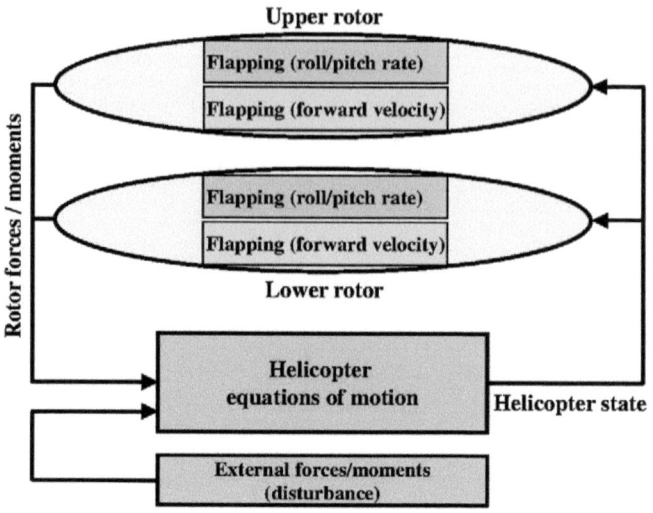

Figure 6.1: Model structure for the parameter search.

helicopter in hover and study its recovery. This disturbance moment is shown in Figure 6.2. It is a rectangular positive roll moment of amplitude 10 Nmm and a duration of 100 ms, which initially causes the helicopter to roll to the right in body-fixed coordinates. Since the main interest of the design parameter study is the passive roll and pitch stabilization, the yaw and altitude degrees of freedom of the helicopter are assumed to be actively controlled. Hence, yaw and heave dynamics are deactivated in the equations of motion of the helicopter.

A very important aspect of the parameter study is the selection of the system parameters to be varied. On the one hand, it is essential to study parameters that actually have an influence on the passive stability behavior. On the other hand, only the variation of parameters that can actually be independently influenced is sensible. Also, in the interest of decent computational time for the parameter search, the set of parameters to be varied should not grow too large, since with every parameter an additional dimension is added to the search space.

The basic set of four parameters that is selected for the parameter search consists of the following:

- Distance between lower rotor and helicopter center of mass d_{lo}.
- Hinge offset e.
- Blade flapping stiffness k_{flap}.
- Blade flapping inertia I_{flap}.

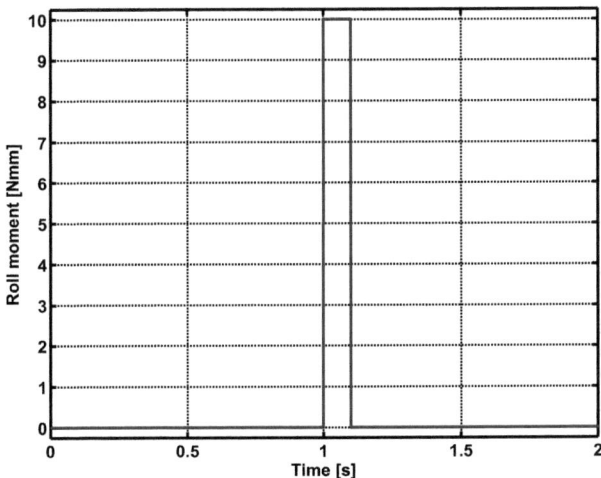

Figure 6.2: Roll disturbance moment applied to the helicopter for the parameter search.

Again, advantage can be taken of Equation (5.3) to reduce the dimension of the search space by one:
$$e_{\text{eqv}} = \frac{g}{\Omega^2 M_{\text{bld}}} k_{\text{flap}}.$$
The blade flapping stiffness k_{flap} is taken into account by equivalent hinge offset e_{eqv}, which is part of the combined hinge offset e_{com} defined in Equation (5.5):
$$e_{\text{com}} = e + e_{\text{eqv}}.$$
Thus, the vector of search parameters $\vec{\Gamma}$ results in
$$\vec{\Gamma} = \begin{bmatrix} d_{\text{lo}} \\ e_{\text{com}} \\ I_{\text{flap}} \end{bmatrix}. \tag{6.1}$$

These parameters offer the advantage that they can be varied without influencing the aerodynamic optimum of the rotors. A variation of the distance between the two rotors could have been interesting from a stability point of view. However, there exists an aerodynamic efficiency optimum for the distance between the two rotors [59], such that its variation for stabilization purposes is not desirable. For the same reason, no aerodynamic parameters of the blade itself are varied.

A criterion needs to be introduced to allow for comparison between several cases of passive stabilization. As it will be shown in the results of Section 6.3, more than one set of parameters leads to passive stabilization. Therefore, for each simulation run a penalty integral is calculated, which accumulates the error in roll and pitch with respect to the hover state of the helicopter. The penalty integral is defined as

$$P = \int_0^{t_{\text{sim}}} \left(|\phi(t)| + |\theta(t)| \right) \, \mathrm{d}t. \tag{6.2}$$

For each simulation run that leads to passive stabilization, the value of the penalty integral P is stored, such that a quantitative comparison between successful parameter sets can be performed.

6.2 Software implementation

Since the complete modular simulation model of Chapter 5 is implemented in Matlab/Simulink, the parameter study is performed using the same software. The parameter search is programmed such that for each search two parameters of $\vec{\Gamma}$ are varied simultaneously. The flow chart of the parameter search implementation is shown in Figure 6.3. For a defined pair of parameters to be varied, first of all the ranges and grid density of these parameters are defined. Then, the simulation model according to Figure 6.1 is run with the first combination of the two parameters. Due to the nature of Simulink, one out of two things will happen: either, the simulation successfully terminates after the predefined simulation time is attained, in which case the calculated value of the penalty integral is simply written to the result matrix. Or, Simulink returns an error statement. This error statement is typically owed to exceeding of the maximally allowable processing time, or because the numerical solver does not converge, which is an indicator for very large derivatives and amplitudes. In that case, the value of `NaN` (**Not a Number**) is written to the result matrix of the penalty integral to facilitate visualization in the data post processing.

After the call of the simulation model, it is checked, whether all parameter combinations for the respective search space have been tried. If that is not the case, the next parameter pair is chosen to recall the simulation, otherwise, the search is terminated and the result matrix filled with numerical values for the cases of stabilization, and `NaN` entries for the cases of destabilization, is ready for postprocessing: when plotted the `NaN`-results leave gaps in the surface plots of the result matrices.

The postprocessing of the parameter search results is straightforward. Due to the use of `NaN` entries for destabilization results, the result matrix can be directly plotted as a function of the two parameter ranges. The result is a surface plot that has gaps or holes for every `NaN` entry. Thus, it can be easily recognized if a parameter combination leads to passive stabilization, and how it compares to other combinations. The surface plots are shown in the following section.

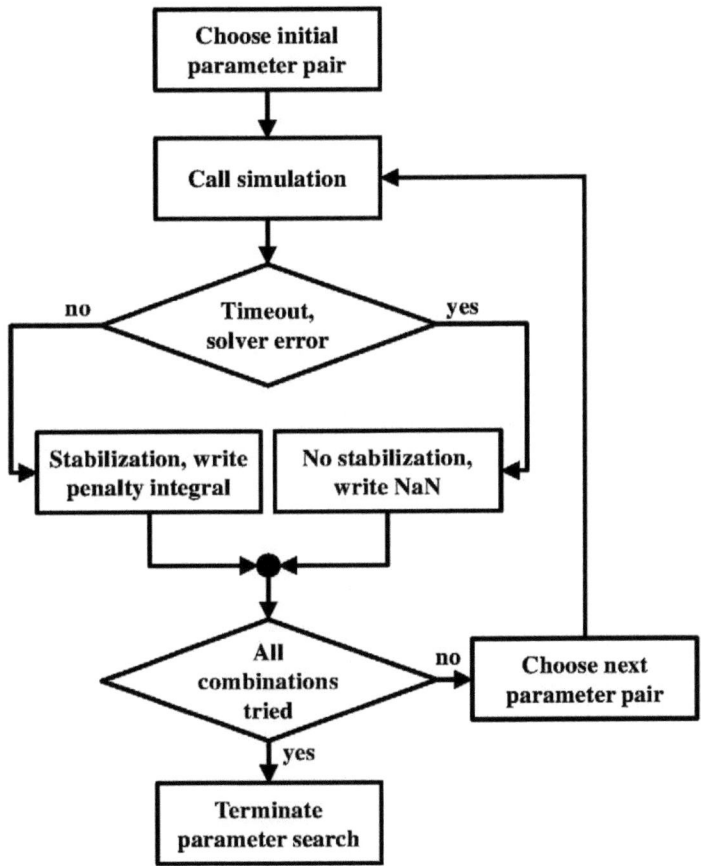

Figure 6.3: Flow chart of the parameter search.

Table 6.1: Design parameter ranges for the variations.

Parameter	Range	Unit	Description
d_{lo}	0 – 0.2	[m]	Lower rotor distance from helicopter CM
e_{com}	0 – 0.07	[m]	Combined hinge offset
I_{flap}	0 – 3×10^{-5}	[kgm^2]	Blade flapping inertia

Table 6.2: Stabilizer bar parameter ranges for the variations.

Parameter	Range	Unit	Description
T_f	0 – 10	[s]	Following time
ϵ_α	-90 – 90	[°]	Phase angle error

6.3 Parameter study results

In this section, several results of the parameter variation are shown, to study the influence of the selected design parameters on the disturbance rejection behavior of the helicopter. Moreover, the simulation framework is used to assess the robustness of the stabilizer bar as a passive stabilization device. For this, an error in the phase angle and the stabilizer bar following time are varied.

6.3.1 Stabilizer bar considerations

Before the parameter study for a passively stable coaxial micro helicopter is considered, the simulation framework is used to assess the stabilization behavior of different parametrizations of the stabilizer bar. The intention is to find the minimal and maximal stabilizer bar following time that destabilizes the helicopter, as it was suspected in Chapter 5, and to investigate, which error in the phase angle α can still be handled until destabilization occurs. For this, the parameters T_f and ϵ_α are varied according to Table 6.2, with a simulation time $t_{sim} = 20$ s and a roll disturbance moment of 30 Nmm. Variation results for the penalty integral P are shown in Figure 6.4 for the full variation domain and in Figure 6.5 for a zoomed view, with the baseline parameters marked as dashed red lines.

The results show an almost symmetric surface of the penalty integral P with respect to a perfectly adjusted stabilizer bar at a phase angle error of $\epsilon_\alpha = 0°$. The surface ends at an approximate phase error of $\epsilon_\alpha = \pm35°$. However, the penalty integral already starts to rise significantly in the range of $\epsilon_\alpha = \pm20°$. Another interesting result is the characteristic gap around $\epsilon_\alpha = 0$ for very high values of the following time. Here, it becomes obvious that a following time chosen too large will destabilize the helicopter, since the stabilizer

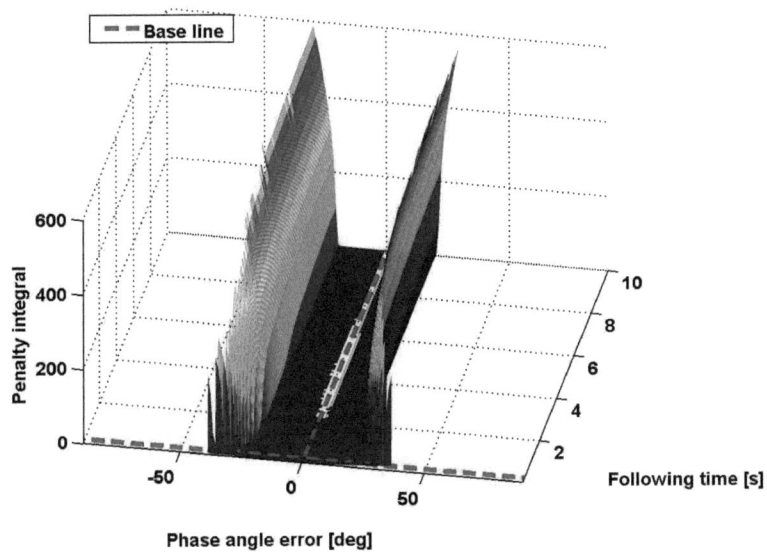

Figure 6.4: Parameter variation result for phase angle error and following time.

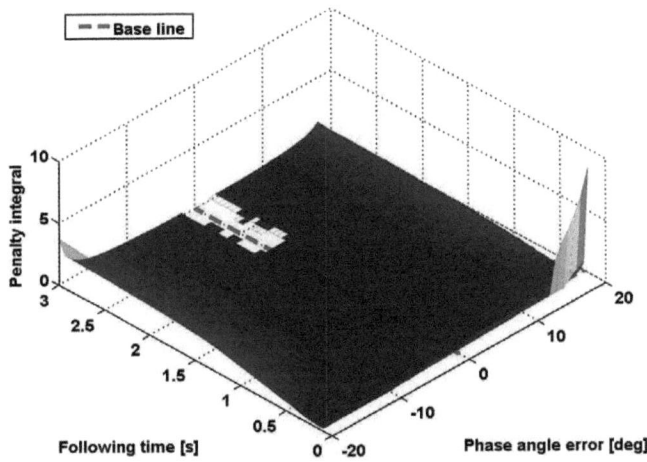

Figure 6.5: Parameter variation result for phase angle error and following time (zoomed).

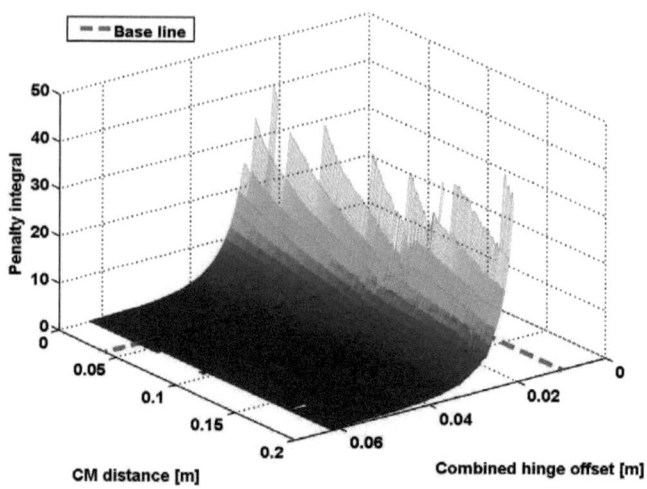

Figure 6.6: Parameter variation result for rotor distance and combined hinge offset (1).

bar is not able to follow the fuselage oscillations any more. Apparently, this problem can be bypassed by introducing a phase angle error, which effectively means that the optimal stabilizer bar angle changes for high values of the following time T_f.

To summarize this result, it seems that the coaxial helicopter tolerates a following time T_f between 100 ms and 2 s and up to approximately 20° in mounting error, before the system is driven into instability.

6.3.2 Rotor distance and combined hinge offset

The results for simultaneous variation of the distance d_{lo} between the lower rotor and the center of mass of the helicopter, and the combined flapping hinge offset e_{com} according to Table 6.1, are shown in Figures 6.6–6.9. The obvious result is that the hinge offset, which corresponds to rotor stiffness, has a significantly higher influence on the stabilization behavior than the rotor distance. While stabilization can be achieved for every value of the rotor distance, for the combined hinge offset this is only possible in the range between approximately 0.025 m and 0.06 m. Also, in accordance with the simulation results in Chapter 5, the base line parameters are located outside the stabilization surface. Finally, it has to be noted that both, too soft and too stiff a rotor can avert passive stabilization. While the transition towards unstable is gradual on the soft side (increase of the penalty integral towards $e_{com} = 0.025$ m), it is very abrupt on the stiff side (towards $e_{com} = 0.061$ m).

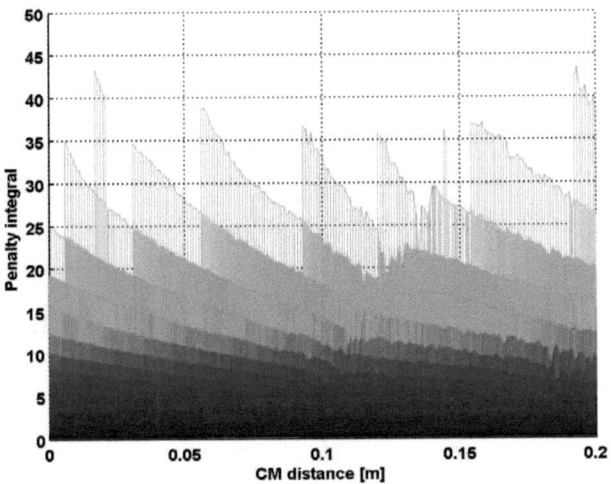

Figure 6.7: Parameter variation result for rotor distance and combined hinge offset (2).

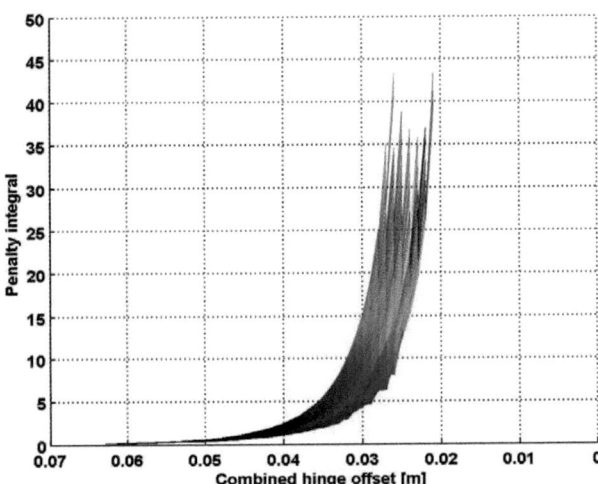

Figure 6.8: Parameter variation result for rotor distance and combined hinge offset (3).

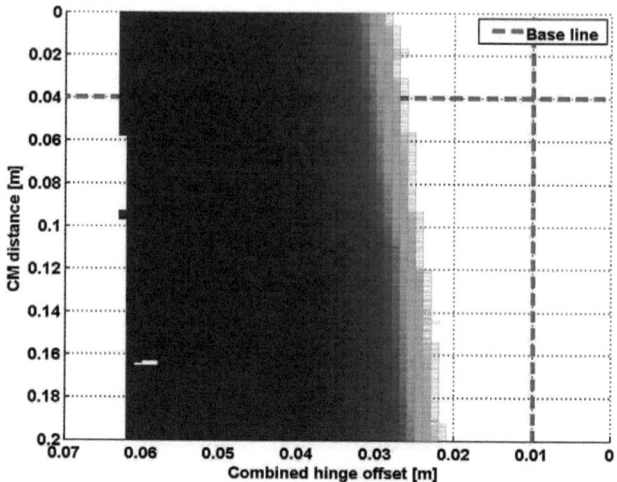

Figure 6.9: Parameter variation result for rotor distance and combined hinge offset (4).

6.3.3 Combined hinge offset and blade flapping inertia

The results for simultaneous variation of the combined hinge offset e_{com}, and the rotor blade flapping inertia I_{flap} according to Table 6.1, are shown in Figures 6.10–6.13. The results look very similar to the previous ones, with the combined hinge offset as the dominating parameter for the passive stabilization. While stabilization can be achieved for every value of the blade flapping inertia, for the combined hinge offset this is only possible in the range between approximately 0.02 m and 0.06 m. With an increase of the blade flapping inertia, the stabilization surface is shifted more towards smaller combined hinge offsets, that is a softer rotor. Also, for a fixed hinge offset, an increase in the blade flapping inertia leads to smaller amplitudes during the stabilization, which is desirable. Finally, in accordance with the results from Chapter 5, the base line parameters lie outside the stabilization surface, which means that for these parameters the helicopter does not recover from the disturbance without a stabilizer bar.

6.3.4 Rotor distance and blade flapping inertia

The results for simultaneous variation of the distance d_{lo} between the lower rotor and the center of mass of the helicopter, and the rotor blade flapping inertia I_{flap} according to Table 6.1, are shown in Figures 6.14–6.17. The stabilization surface is larger than for the previous results, implying that passive stabilization is relatively insensitive to the two

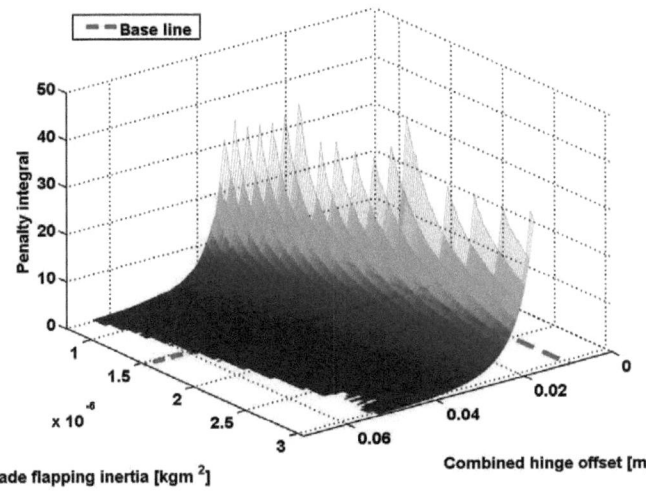

Figure 6.10: Parameter variation result for combined hinge offset and blade flapping inertia (1).

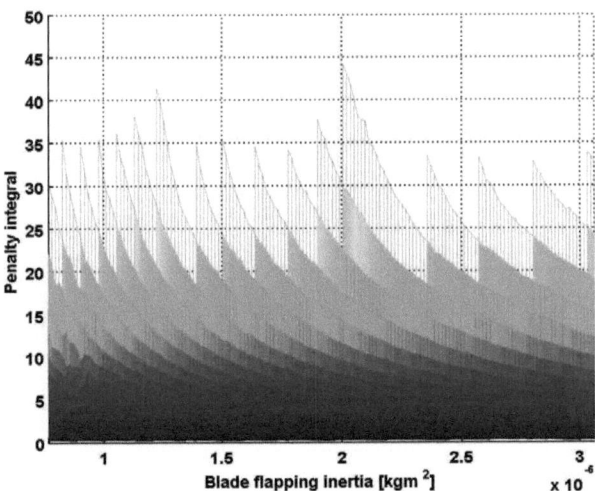

Figure 6.11: Parameter variation result for combined hinge offset and blade flapping inertia (2).

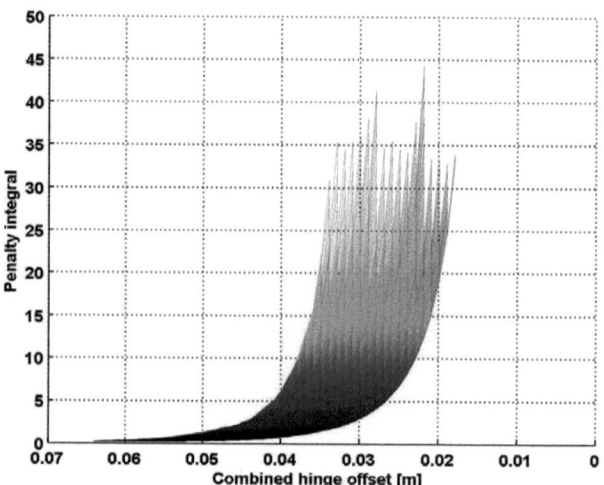

Figure 6.12: Parameter variation result for combined hinge offset and blade flapping inertia (3).

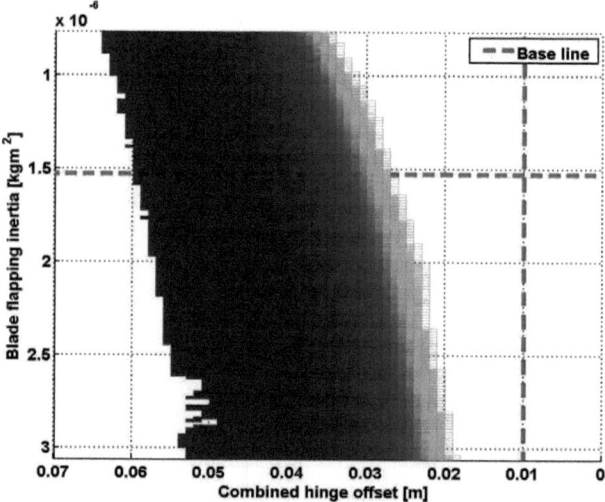

Figure 6.13: Parameter variation result for combined hinge offset and blade flapping inertia (4).

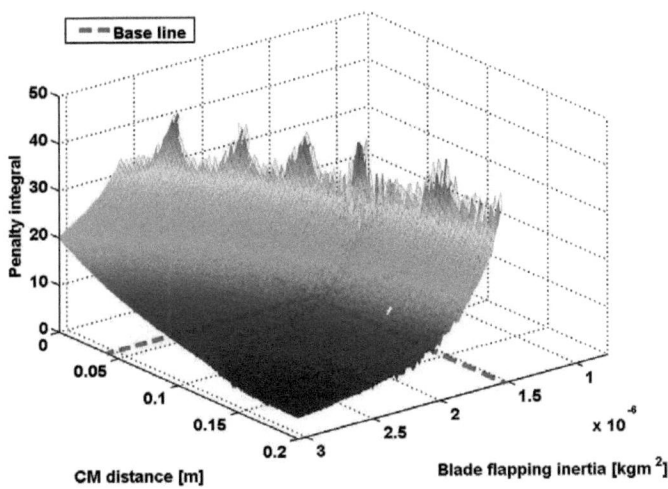

Figure 6.14: Parameter variation result for rotor distance and blade flapping inertia (1).

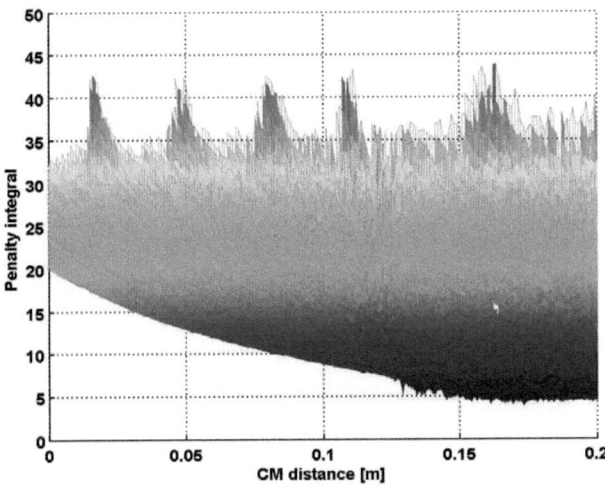

Figure 6.15: Parameter variation result for rotor distance and blade flapping inertia (2).

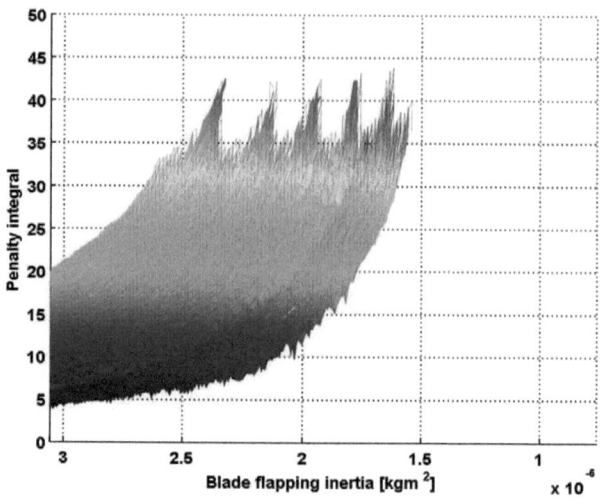

Figure 6.16: Parameter variation result for rotor distance and blade flapping inertia (3).

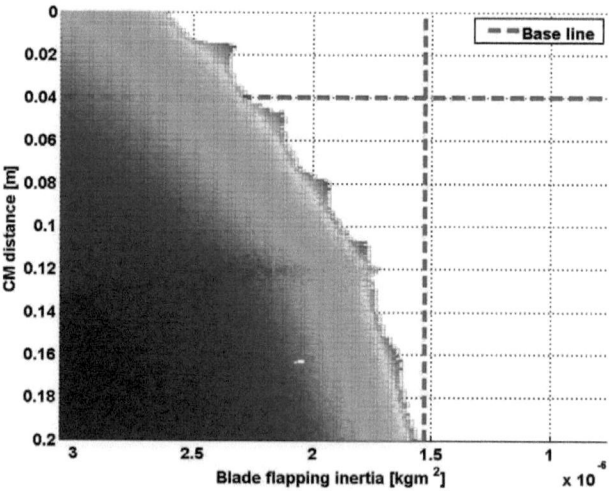

Figure 6.17: Parameter variation result for rotor distance and blade flapping inertia (4).

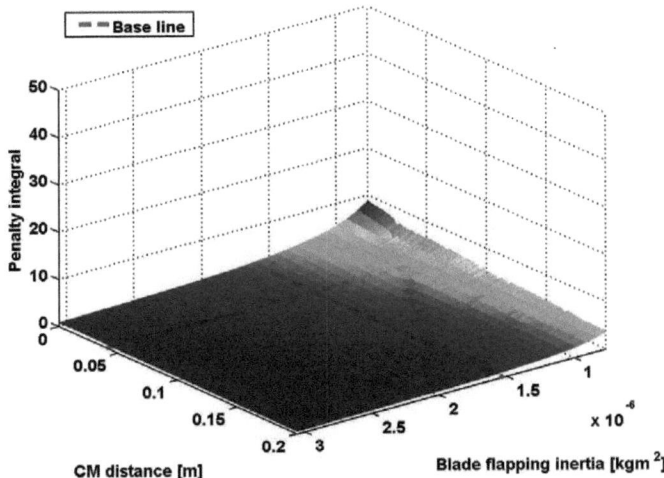

Figure 6.18: Final parameter variation result for rotor distance and blade flapping inertia with $e_{\text{com}} = 0.04$ (1).

parameters varied. What can be observed is that an increase in the distance between lower rotor and center of mass allows for more freedom in the selection of the flapping inertia. Nevertheless, the chosen base line parameters lie outside of the stabilization surface. This is however, owed to the selection of the combined hinge offset e_{com}, as will be shown in the results of the following section.

6.3.5 Parameter selection

The previous result sections have shown that the combined hinge offset e_{com} is the most important design parameter for the passive stabilization, while rotor distance d_{lo} and flapping inertia I_{flap} are less relevant. Therefore, the parameter variation of d_{lo} and I_{flap} is rerun with a more appropriate value of the combined hinge offset, namely $e_{\text{com}} = 0.04\,\text{m}$. The results of this variation are shown in Figures 6.18–6.21. For the new value of $e_{\text{com}} = 0.04\,\text{m}$, the result surface covers the complete parameter variation domain, which means that for all combinations of rotor distance d_{lo} and blade flapping inertia I_{flap} passive stabilization is achieved. This includes the baseline parameter values of the rotor distance and the blade flapping inertia. Moreover, the average value of the penalty integral P is much lower than for all previous parameter variations. Hence, the set of design parameters chosen for a passively stable helicopter are

- $d_{\text{lo}} = 0.04\,\text{m}$,

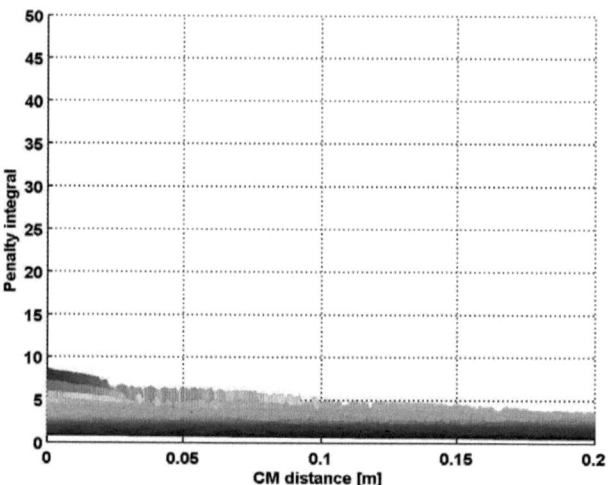

Figure 6.19: Final parameter variation result for rotor distance and blade flapping inertia with $e_{\text{com}} = 0.04$ (2).

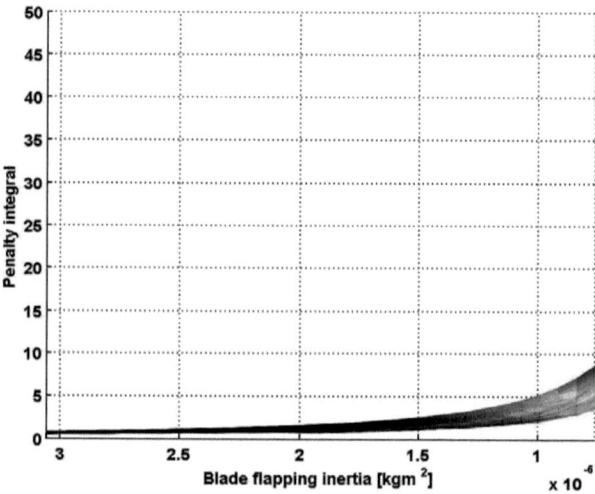

Figure 6.20: Final parameter variation result for rotor distance and blade flapping inertia with $e_{\text{com}} = 0.04$ (3).

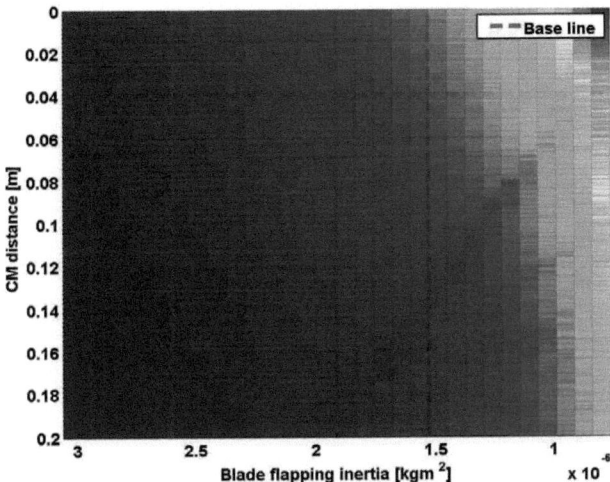

Figure 6.21: Final parameter variation result for rotor distance and blade flapping inertia with $e_{\text{com}} = 0.04$ (4).

- $e_{\text{com}} = 0.04\,\text{m}$,
- $I_{\text{flap}} = 1.5 \times 10^{-6}\,\text{kgm}^2$.

With these parameters, it is theoretically possible to build a coaxial micro helicopter that is passively stable in roll and pitch simply by selection of its design parameters. Additional devices like a stabilizer bar or a Proxflyer rotor system are not required.

6.4 Summary

In this chapter, a design parameter study is presented. It aims at finding design parameters for a coaxial micro helicopter that is stable in roll and pitch without using active control or additional passive devices like the stabilizer bar or the Proxflyer rotor system.

The parameter study is based on a reduced version of the modular dynamic simulation model in Chapter 5. All active control inputs and the stabilizer bar module are removed, for the simulation runs the system is initially at hover. Heave and yaw dynamics are assumed to be stabilized by feedback control and hence disabled for simulation. The system is then subjected to a standard roll disturbance, which is strong enough to destabilize the system at baseline parameter values. Simulations are run for varying combinations of the rotor distance d_{lo}, the combined hinge offset e_{com} and the blade flapping inertia I_{flap}, for each run two parameters are varied simultaneously, while the third one is held at its base line value. Self-stabilization is evaluated in terms of a penalty integral.

The simulation results show that for each variation, parameter combinations can be found that lead to passive stabilization. The most obvious result is that the combined hinge offset e_{com} is the most sensitive parameter for passive roll and pitch stabilization, while the rotor distance d_{lo} and the flapping inertia I_{flap} put barely any limitations on the stabilizing parameter domains. Hence, for a passively stable design, it is most crucial to select the combined hinge offset carefully. It represents the actual hinge offset and the blade flapping stiffness, so in principle the design of the rotor blade root. With the parameters found in this study, blades for a passively stable helicopter can be produced in future work.

A side result of the simulation framework introduced in this chapter is a parameter study for the governing parameters of the stabilizer bar, namely following time and phase angle. Here, it can be shown that the stabilizer bar is relatively tolerant to a mounting error. Even if mounted around 20° off the optimal value, the system can still be stabilized. Moreover, apart from destabilization due to too small a following time, also too large a following time will destabilize the system. In that case, the stabilizer bar becomes sluggish and too slow to follow the fuselage oscillations. As a result, the correcting influence of the stabilizer becomes the opposite and amplifies fuselage oscillations, leading to destabilization.

Chapter 7

Conclusion

In this work, several contributions to the design of autonomous micro helicopters in coaxial rotor configuration are made. They have been achieved within the European framework project muFly, with the goal to build an autonomous micro helicopter that is comparable to a small bird in size and mass.
In the first part of this thesis, two prototypes for the muFly project helicopter are introduced, which are both aimed at full autonomy, but follow different design paradigms: the first prototype is designed in a modular way to allow for easy exchange and testing of hardware components, as well as control algorithm testing. However, this high degree of modularity leads to a total mass that consumes most of the thrust margin, making it impossible to carry the additional mass of the x-y-position sensor consisting of laser diodes and omnidirectional camera. Particular reasons are the high structural and electronics mass of the helicopter. The second prototypes follows the design paradigm of utmost integration of all components, including the x-y-position sensor, which makes hardware component exchange difficult, but leads to a drastic reduction of the total mass of the helicopter. The high degree of integration is achieved by dual use of the system electronics as structural components, such that almost no additional structural mass has to be expended. Also, by advantageous layout of the electronic boards, most of the cables and connectors can be cut down. Moreover, the interlacing design of the electronic boards leads to a very stiff frame, which guarantees for the necessary mounting precision of lasers and camera with respect to each other. The total mass saving on the second prototype accounts for approximately 17 % with respect to the first prototype, while the payload mass percentage has been increased.
In the second part of this thesis, a theoretical foundation for the design assessment of future coaxial micro helicopter prototypes is laid. First of all, the stabilizer bar, which is a passive stabilization device for the roll and pitch rates of the helicopter, is investigated in dynamic simulations and experiments. In order to have precise values for the stabilizer bar following time, which is used later in the modular dynamic model, and to find the correct phase angle of the muFly rotor system, which is important for the steering input

and stabilizer bar phasing on the helicopter prototypes, a multi-body dynamic model of the stabilizer bar is established and compared to experimental results measured on a test rig. The results of this investigation show a good match between simulation and experiments for both the following time and the optimal phase angle. While the following time depends more on the flapping inertia of the stabilizer bar than on the rotor speed, the phase angle is principally independent of the parameters of the stabilizer bar and depends on the rotor speed and design parameters of the hingeless rotor.

Following the investigation of the stabilizer bar, this thesis introduces a modular simulation model for coaxial micro helicopters. It is based on first principles and takes into account the passive rotor blade flapping dynamics resulting from roll, pitch and forward velocities of the helicopter, as well as the active flapping contribution from cyclic blade pitch changes. Moreover, the model comprises a stabilizer module and two different modules for the most prevalent steering mechanisms on micro helicopters, which are swash plate steering and steering by displacement of the center of mass. With these modules, different configurations of the coaxial helicopter can be simulated and quantitatively evaluated to support design decisions. The simulation results show the advantageous characteristics of the coaxial rotor configuration in terms of force and moment disturbance rejection: while the restoring forces and moments are twice as large as for a single main rotor configuration, the cross axis forces and moments are almost perfectly canceled out due to the opposite directions of rotation of the rotors. However, this behavior is also a disadvantage in terms of the steering performance, which is decreased by the rejection of the active steering inputs of the rotors. Based on these results, the coaxial rotor configuration is certainly a good choice for MAVs that need a high degree of stability and spend most of their flight time in hover and slow translations, but a bad choice for highly dynamic MAVs that are intended to perform highly dynamic flight maneuvers.

The evaluation of the steering mechanisms, which are swash plate steering and center of mass displacement steering, shows a clear advantage for the swash plate: it delivers larger steering amplitudes, has a higher stroke reserve and consumes approximately three times less energy per stroke. Hence, swash plate mechanisms should be used on micro helicopters whenever possible. The only situation in favor of mass displacement steering could be a helicopter that is so small that a swash plate cannot be realized due to the limits of miniaturization.

Simulation results of the stabilizer bar module show its stabilizing behavior in roll and pitch, but also that this stabilization appears between limits: if the stabilizer bar inertia is too low, it follows the helicopter motion too quickly to effect a sufficiently large restoring moment, if the inertia is too high, it is not able to follow oscillations of the fuselage quickly enough, which results in a phase shift between the upper and lower rotor reaction moments and subsequently in destabilization.

Finally, the introduced modular simulation model is used to study a new approach to passive roll and pitch stability by means of a design parameter study. The goal of the study is to achieve passive roll and pitch stabilization of the helicopter without an additional device like the stabilizer bar, which introduces additional mass in the system and drag and inertia on the rotor, which in turn leads to an increased energy consumption and reduced flight endurance. Hence, in the study the design parameters that influence the steering moments of the helicopter are varied, i.e. the combined hinge offset, blade flapping inertia and distance between helicopter center of mass and lower rotor. The results of this study show that passive stabilization by advantageous selection of these parameters is possible, and that the key parameter for this is the combined hinge offset, while flapping inertia and the distance between lower rotor and center of mass of the helicopter can be chosen relatively freely.

Future designs of autonomous micro helicopters will certainly follow a highly integrated design approach to keep their total mass as low as possible. Since the first step towards an autonomous vehicle is energy autonomy, reasonable mission flight times of at least ten minutes can only be achieved by minimal mass of the helicopter, together with improved batteries. Moreover, these helicopters will be steered by swash plate mechanisms. Finally, the stabilizer bar, which is undoubtedly a simple and easy means of passive helicopter stabilization, will be sacrificed for the goal of reduced power consumption. There are existing examples of successful closed loop control of coaxial micro helicopters without a stabilizer bar, and the required hardware and control algorithms have to be on board an autonomous micro helicopter either way. If passive roll and pitch stabilization of the helicopter is the goal, it could be achieved by the rotor system modifications proposed in this thesis.

Appendix A

Historical evolution of coaxial helicopters

Although coaxial helicopters are rather rarely seen in the skies and clearly outnumbered by helicopters with the conventional main and tail rotor, the coaxial configuration has a long history and was already envisioned by early dreamers and helicopter pioneers. This section gives a brief historical overview and time line of the evolution of coaxial helicopters; the contents are based on the results of [53]. The designs and types mentioned throughout the section are shown in an time line in Figure A.9 for the early days and in Figure A.10 for the modern ages of the coaxial helicopter.

Already in 1493, Leonardo Da Vinci envisioned aerial vehicles with hovering capability and the possibility to take off and land vertically by using a large airscrew. The basic idea of the helicopter was born. In the 18th century, first designs with a set of two coaxial rotors became known. These were the Aerodynamic (Figure A.1 (left)) by Lomonosov in 1754, which had a gearbox to achieve opposite rotation of the rotors, and a toy by Launoy and Bienvenu in 1784 (Figure A.1 (center)), which effected unsynchronized counter rotation by means of a rotational spring. Although pioneer work, these designs were rather early studies than usable helicopters. The first patent for a coaxial helicopter design was awarded by the British Patent Office to Henry Bright in 1859 [32]. Interestingly, his design (Figure A.1 (right)) seems to incorporate an anti-torque device in the form of a tail fin,

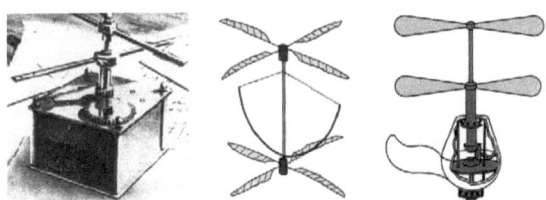

Figure A.1: Lomonosov's Aerodynamic (left), a toy by Launoy and Bienvenu (center), and Henry Bright's patent (right).

Figure A.2: The Breguet-Richet Gyroplane No.1.

Figure A.3: Sikorsky's S-1 (left) and Antonov's Helicoplane (right).

despite the coaxial rotor configuration. This is an indicator that at that point the benefit of a coaxial rotor configuration was not fully understood. Hence, it was not until fall 1907 that the first man carrying helicopter went to flight: the Breguet-Richet Gyroplane No.1 (Figure A.2) achieved manned level flight at an altitude of about 60 cm. The configuration is principally a quadrotor, but employs four coaxial rotor sets for a total of eight rotors. For its time it was a design with immense mechanical complexity, since all rotors were driven by a single engine via a belt transmission system. Nowadays, the coaxial-multirotor configuration has become interesting again in the UAV and MAV domain [6]. In Russia, helicopter pioneer Igor I. Sikorsky and military engineer K.A. Antonov (not to be confused with the airplane designer O.K. Antonov) also developed coaxial helicopter prototypes in 1909 (S-1 by Sikorsky, Figure A.3 (left)) and 1910 (Helicoplane by Antonov, Figure A.3 (right)), respectively. These proved to be functioning mechanisms, but were not able to fly due to a lack of engine power. While the helicopter development in general did not see any progress in World War I, it rapidly gained momentum after the war. In the 1920s, Raúl Pateras Pescara, an Argentinian aeronautics pioneer, filed numerous patents on coaxial designs and had his most successful flights in 1924 with the Pescara No.3 (Figure A.4 (left)). This coaxial helicopter achieved several flights with durations of more than ten minutes and set a helicopter speed record of 13 km/h. It also employed mechanisms for collective and cyclic blade pitch control, which was a remarkable innovation for that time. In 1930, a coaxial helicopter built by Corradino D'Ascanio (Figure A.4 (right)) set three Fédération Aéronautique Internationale (FAI) records. What is interesting about his design is the fact that he used a trailing edge flap mechanism to vary the angle of attack

Figure A.4: The Pescara No.3 (left), and D'Asciano's record helicopter (right).

Figure A.5: The Breguet G.IIE Gyroplane (left), and Manzolini's Libullella (right).

of each blade, a concept that was unmanageable on commercial helicopters for a long time due to actuator and material restrictions. It has, however, become newly interesting in recent times, when better materials and actuators for helicopters became available [57]. After this period of the first working prototypes, the time had come for real performers. In 1935, French aeronautics pioneer Louis C. Breguet sent the Gyroplane Laboratoire to first flight. The design, which was one of the first practical helicopters, incorporated cyclic and collective blade pitch control and set FAI records for altitude (158 m) flight endurance (1:02 h) and flight speed (120 km/h). The following years, especially after World War II, saw further acceleration of the coaxial helicopter evolution. Various designs were introduced on both sides of the Atlantic. United States helicopter pioneers Stanley Hiller and Arthur Young with Bell Helicopter released coaxials in 1944 (Hiller XH-44) and 1945 (Bell Coaxial Rotor Helicopter), another manufacturer in the US was Brantly with their rather unsuccessful B-1 (1946). Between 1947 and 1952, the Gyrodyne company released three coaxial designs that were partially enhanced by outrigged propellers for a higher forward speed. On the European side, the Breguet G.IIE Gyroplane (1949, Figure A.5 (left)) and the Libullella by Ettore Manzolini (1952, Figure A.5 (right)) were introduced. A chapter of its own is the Kamov company, which remains the most significant producer of coaxial helicopters until today. The company started in the coaxial domain in 1947 with the Ka-8, which was basically a flying motorcycle, having no closed cabin and a makeshift pilot seat. The Ka-10 in 1949 introduced better flight characteristics and reliability, followed by the very compact Ka-15, which was especially designed for ship-based operations on narrow flight decks. Driven by the need for air transport capacity in the former Soviet Union and by the Cold War, a series of cargo and utility helicopters (Ka-15 – Ka-32) as well as combat helicopters (Ka-29 and Ka-50) were released. Especially the combat versions of

Figure A.6: Utility helicopter Kamov Ka-32 (left) and combat helicopter Kamov Ka-50 (right).

Figure A.7: Coaxial helicopter UAVs: the Gyrodyne QH-50 (left) and the Kamov Ka-37 (right).

the coaxial configuration did convince by supreme maneuverability and survivability. The possibility of high yaw rates without driving a tail rotor into the dangerous vortex ring state [95] accounted for a large maneuverability advantage in combat. In 1997, already after the collapse of the Soviet Union, a more modern combat helicopter was introduced with the Ka-52. An extensive overview over the Soviet coaxial helicopter developments is given in [44]. Naturally, due to its stable flight characteristics and its compactness, the coaxial helicopter has been quickly identified as an ideal configuration for rotary wing UAVs. Hence, it is not surprising that the first helicopter UAV, the Gyrodyne QH-50 of 1961 [67] (see Figure A.7 (left)), was a coaxial. Its intended use was for anti-submarine missions from narrow ship decks. Another noteworthy development is the Ka-37 from 1993 (Figure A.7 (right)), a joint development of Kamov and the Daewoo company, which can be considered as one of the first helicopter UAVs for civilian purposes such as aerial photography. The coaxial rotor configuration is still within the focus of research and development. Especially in unconventional configurations, the coaxial principle can offer advantages over the conventional rotor configuration. In this context the early Gyrodyne coaxial designs with additional forward propellers currently undergo a renaissance: the coaxial research design Sikorsky X2, which comprises a pusher propeller for better forward flight characteristics and higher forward flight speed, shows that the coaxial rotor configuration is not yet at the limits of its performance capabilities (see Figure A.8). Hence, the future will most likely bring a number of unconventional, yet high-performing helicopters with coaxial rotors.

Figure A.8: Early Gyrodyne prototype GCA2 (left) and future design Sikorsky X2 (right).

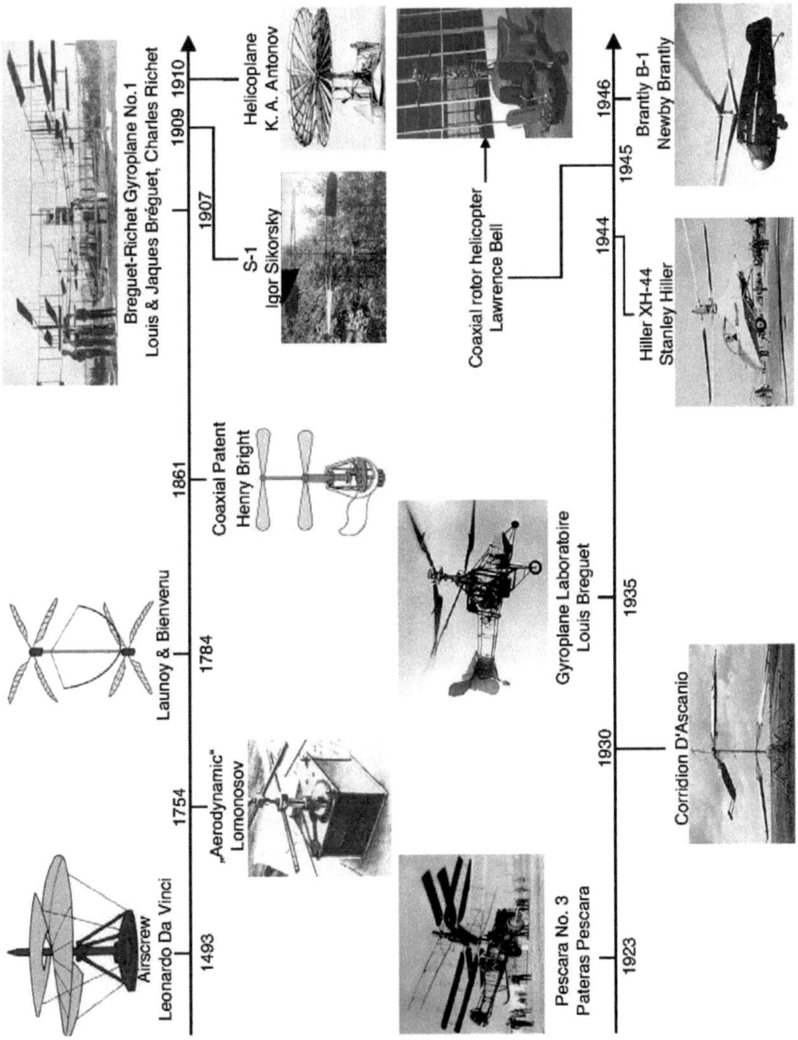

Figure A.9: Timeline for the early days of the coaxial helicopter.

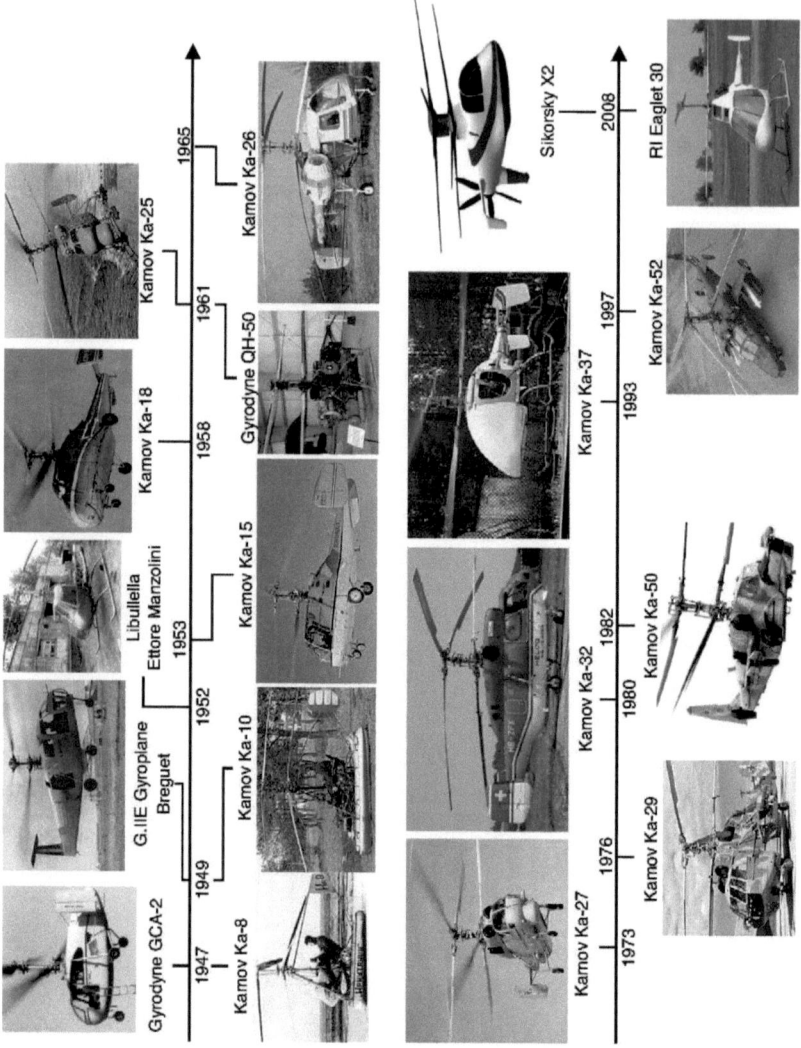

Figure A.10: Timeline for the modern ages of the coaxial helicopter.

Appendix B

Matrices for the stabilizer bar model

B.1 Transformation matrices

The elementary rotational transformation matrices $\mathbf{R}_x(\alpha)$, $\mathbf{R}_y(\beta)$ and $\mathbf{R}_z(\gamma)$ are defined as follows [37]. They are in accordance with the right hand rule, where the rotation $\mathbf{R}_x(\alpha)$ rotates the y-axis towards the z-axis, the rotation $\mathbf{R}_y(\beta)$ rotates the z-axis towards the x-axis and the rotation $\mathbf{R}_z(\gamma)$ rotates the x-axis towards the y-axis.

$$\mathbf{R}_x(\alpha) = \begin{bmatrix} 1 & 0 & 0 \\ 0 & \cos\alpha & -\sin\alpha \\ 0 & \sin\alpha & \cos\alpha \end{bmatrix},$$

$$\mathbf{R}_y(\beta) = \begin{bmatrix} \cos\beta & 0 & \sin\beta \\ 0 & 1 & 0 \\ -\sin\beta & 0 & \cos\beta \end{bmatrix},$$

$$\mathbf{R}_z(\gamma) = \begin{bmatrix} \cos\gamma & -\sin\gamma & 0 \\ \sin\gamma & \cos\gamma & 0 \\ 0 & 0 & 1 \end{bmatrix}.$$

B.2 Linearized system matrices

The non-zero elements of the linearized stabilizer bar model matrices $\mathbf{M}(t)$, $\mathbf{D}(t)$ and $\mathbf{C}(t)$ are given in the following:

B.2.1 Mass matrix $\mathbf{M}(t)$

$$M_{11} = 2I_{\text{bld},yy} \sin^2 \psi + 2I_{\text{bld},xx} \cos^2 \psi + 2m_{\text{bld}} \sin^2 \psi \, (d_{\text{cm}} + e)^2$$
$$+ I_{\text{sb},xx} \cos^2 (\psi + \alpha) + I_{\text{sb},yy} \sin^2 (\psi + \alpha) + I_{\text{fus},xx} + h^2 (2m_{\text{bld}} + m_{\text{sb}})$$

$$M_{12} = M_{21} = -2m_{\text{bld}} \, (d_{\text{cm}} + e)^2 \sin \psi \cos \psi + 2 \, (I_{\text{bld},xx} - I_{\text{bld},yy}) \sin \psi \cos \psi$$
$$+ (I_{\text{sb},xx} - I_{\text{sb},yy}) \sin (\psi + \alpha) \cos (\psi + \alpha)$$

$$M_{13} = M_{31} = 2\kappa I_{\text{bld},xx} \cos \psi - I_{\text{sb},yy} \sin (\psi + \alpha)$$

$$M_{14} = M_{41} = -\sin \psi \, [m_{\text{bld}} d_{\text{cm}} (d_{\text{cm}} + e) + I_{\text{bld},yy}]$$

$$M_{15} = M_{51} = \sin \psi \, [m_{\text{bld}} d_{\text{cm}} (d_{\text{cm}} + e) + I_{\text{bld},yy}]$$

$$M_{22} = 2I_{\text{bld},xx} \sin^2 \psi + 2 \cos^2 \psi \left[m_{\text{bld}} \, (d_{\text{cm}} + e)^2 + I_{\text{bld},yy} \right]$$
$$+ I_{\text{sb},xx} \sin^2 (\psi + \alpha) + I_{\text{sb},yy} \cos^2 (\psi + \alpha)$$
$$+ h^2 (2m_{\text{bld}} + m_{\text{sb}}) + I_{yy,\text{fus}}$$

$$M_{23} = M_{32} = 2I_{\text{bld},xx} \kappa \sin \psi + I_{\text{sb},yy} \cos (\psi + \alpha)$$

$$M_{24} = M_{42} = \cos \psi \, [m_{\text{bld}} d_{\text{cm}} (d_{\text{cm}} + e) + I_{\text{bld},yy}]$$

$$M_{25} = M_{52} = -\cos \psi \, [m_{\text{bld}} d_{\text{cm}} (d_{\text{cm}} + e) + I_{\text{bld},yy}]$$

$$M_{33} = I_{\text{sb},yy} + 2\kappa^2 I_{\text{bld},xx}$$

$$M_{44} = I_{\text{bld},yy} + m_{\text{bld}} d_{\text{cm}}^2$$

$$M_{55} = I_{\text{bld},yy} + m_{\text{bld}} d_{\text{cm}}^2$$

B.2.2 Damping matrix $\mathbf{D}(t)$

$D_{11} = 2\sin\psi\cos\psi\left[I_{\text{bld},yy} - 2I_{\text{bld},xx} + 2m_{\text{bld}}\left(d_{\text{cm}} + e\right)^2\right]$
$\qquad + 2\sin\left(\psi + \alpha\right)\cos\left(\psi + \alpha\right)\left(I_{\text{sb},yy} - I_{\text{sb},xx}\right)$

$D_{12} = 2\cos^2\left(\psi + \alpha\right)\left(I_{\text{sb},xx} - I_{\text{sb},yy}\right) + 4\cos^2\psi\left(I_{\text{bld},xx} - I_{\text{bld},yy}\right)$
$\qquad + 4m_{\text{bld}}\sin^2\psi\left(d_{\text{cm}} + e\right)^2 - I_{\text{sb},xx} + I_{\text{sb},yy}$
$\qquad + I_{\text{sb},zz} + I_{\text{fus},zz} - 2\left(I_{\text{bld},xx} - I_{\text{bld},yy} - I_{\text{bld},zz}\right)$

$D_{13} = -2\kappa\sin\psi\left(I_{\text{bld},xx} + I_{\text{bld},yy} - I_{\text{bld},zz}\right) + \cos\left(\psi + \alpha\right)\left(I_{\text{sb},xx} + I_{\text{sb},yy} - I_{\text{sb},zz}\right)$

$D_{14} = -\cos\psi\left(I_{\text{bld},yy} + I_{\text{bld},xx} - I_{\text{bld},zz}\right)$

$D_{15} = \cos\psi\left(I_{\text{bld},yy} + I_{\text{bld},xx} - I_{\text{bld},zz}\right)$

$D_{21} = -2\left[\left(I_{\text{sb},yy} - I_{\text{sb},xx}\right)\cos^2\left(\psi + \alpha\right) - 4\cos^2\psi\left(I_{\text{bld},xx} + I_{\text{bld},yy}\right)\right.$
$\qquad + 4m_{\text{bld}}\cos^2\psi\left(d_{\text{cm}} + e\right)^2$
$\qquad \left. + 2\left(I_{\text{bld},xx} - 2I_{\text{bld},yy} + I_{\text{bld},zz}\right)I_{\text{sb},xx} - I_{\text{sb},yy} + I_{\text{sb},zz} + I_{\text{bld},zz}\right]$

$D_{22} = -2\left[2m_{\text{bld}}\sin\psi\cos\psi\left(d_{\text{cm}} + e\right)^2 - 2\sin\psi\cos\psi\left(I_{\text{bld},xx} + I_{\text{bld},yy}\right)\right.$
$\qquad \left. - \sin\left(\psi + \alpha\right)\cos\left(\psi + \alpha\right)\left(I_{\text{sb},xx} + I_{\text{sb},yy}\right)\right]$

$D_{23} = 2\kappa\cos\psi\left(I_{\text{bld},xx} + I_{\text{bld},yy} - I_{\text{bld},zz}\right) - \sin\left(\psi + \alpha\right)\left(I_{\text{sb},xx} + I_{\text{sb},yy} - I_{\text{sb},zz}\right)$

$D_{24} = -\sin\psi\left(I_{\text{bld},yy} + I_{\text{bld},xx} - I_{\text{bld},zz}\right)$

$D_{25} = \sin\psi\left(I_{\text{bld},yy} + I_{\text{bld},xx} - I_{\text{bld},zz}\right)$

$D_{31} = -2\kappa\sin\psi\left(I_{\text{bld},xx} + I_{\text{bld},yy} - I_{\text{bld},zz}\right) + \cos\left(\psi + \alpha\right)\left(I_{\text{sb},xx} + I_{\text{sb},yy} - I_{\text{sb},zz}\right)$

$D_{32} = 2\kappa\cos\psi\left(I_{\text{bld},xx} - I_{\text{bld},yy} + I_{\text{bld},zz}\right) + \sin\left(\psi + \alpha\right)\left(I_{\text{sb},xx} - I_{\text{sb},yy} - I_{\text{sb},zz}\right)$

$D_{34} = -\kappa\left(I_{\text{bld},xx} + I_{\text{bld},yy} - I_{\text{bld},zz}\right)$

$D_{35} = \kappa\left(I_{\text{bld},xx} + I_{\text{bld},yy} - I_{\text{bld},zz}\right)$

$D_{41} = -\cos\psi\left[2m_{\text{bld}}d_{\text{cm}}\left(d_{\text{cm}} + e\right) - I_{\text{bld},xx} + I_{\text{bld},yy} + I_{\text{bld},zz}\right]$

$D_{42} = -\sin\psi\left[2m_{\text{bld}}d_{\text{cm}}\left(d_{\text{cm}} + e\right) - I_{\text{bld},xx} + I_{\text{bld},yy} + I_{\text{bld},zz}\right]$

$D_{43} = \kappa\left(I_{\text{bld},xx} + I_{\text{bld},yy} - I_{\text{bld},zz}\right)$

$D_{51} = \cos\psi\left[2m_{\text{bld}}d_{\text{cm}}\left(d_{\text{cm}} + e\right) - I_{\text{bld},xx} + I_{\text{bld},yy} + I_{\text{bld},zz}\right]$

$D_{52} = \sin\psi\left[2m_{\text{bld}}d_{\text{cm}}\left(d_{\text{cm}} + e\right) - I_{\text{bld},xx} + I_{\text{bld},yy} + I_{\text{bld},zz}\right]$

$D_{53} = -\kappa\left(I_{\text{bld},xx} + I_{\text{bld},yy} - I_{\text{bld},zz}\right)$

B.2.3 Spring matrix $\mathbf{C}(t)$

$$C_{13} = -2\kappa \cos\psi \left(I_{\text{bld},yy} - I_{\text{bld},zz}\right) + \sin\left(\psi + \alpha\right)\left(I_{\text{sb},xx} - I_{\text{sb},zz}\right)$$

$$C_{14} = \sin\psi \left[I_{\text{bld},xx} - I_{\text{bld},zz} - m_{\text{bld}}d_{\text{cm}}\left(d_{\text{cm}} + e\right)\right]$$

$$C_{15} = -\sin\psi \left[I_{\text{bld},xx} - I_{\text{bld},zz} - m_{\text{bld}}d_{\text{cm}}\left(d_{\text{cm}} + e\right)\right]$$

$$C_{23} = -2\kappa \sin\psi \left(I_{\text{bld},yy} - I_{\text{bld},zz}\right) - \cos\left(\psi + \alpha\right)\left(I_{\text{sb},xx} - I_{\text{sb},zz}\right)$$

$$C_{24} = -\cos\psi \left[I_{\text{bld},xx} - I_{\text{bld},zz} - m_{\text{bld}}d_{\text{cm}}\left(d_{\text{cm}} + e\right)\right]$$

$$C_{25} = \cos\psi \left[I_{\text{bld},xx} - I_{\text{bld},zz} - m_{\text{bld}}d_{\text{cm}}\left(d_{\text{cm}} + e\right)\right]$$

$$C_{33} = -2\kappa^2 \left(I_{\text{bld},yy} - I_{\text{bld},zz}\right) - I_{\text{sb},xx} + I_{\text{sb},zz}$$

$$C_{44} = \frac{k_{\text{flap}}}{\Omega^2} - I_{\text{bld},xx} + I_{\text{bld},zz} + m_{\text{bld}}d_{\text{cm}}\left(d_{\text{cm}} + e\right)$$

$$C_{55} = \frac{k_{\text{flap}}}{\Omega^2} - I_{\text{bld},xx} + I_{\text{bld},zz} + m_{\text{bld}}d_{\text{cm}}\left(d_{\text{cm}} + e\right)$$

Bibliography

[1] Autonomous Intelligent Systems AIS, ALU Freiburg.
http://ais.informatik.uni-freiburg.de/, (12/19/2009).

[2] Autonomous Systems Lab, ETH Zürich.
http://www.asl.ethz.ch, (12/19/2009).

[3] BeCAP TU Berlin.
http://www.becap.tu-berlin.de/, (12/19/2009).

[4] CEDRAT Technologies.
http://www.cedrat.com, (12/19/2009).

[5] CSEM.
http://www.csem.ch/, (12/19/2009).

[6] Draganfly Helicopters.
http://www.draganfly.com, (12/19/2009).

[7] Epson.
http://www.epson.co.jp/e/newsroom/news_2004_08_18.htm, (12/19/2009).

[8] Project muFly.
http://www.mufly.org, (12/19/2009).

[9] Walkera Helicopters.
http://www.walkera.com, (12/19/2009).

[10] WES-Technik BLDC motors.
http://www.wes-technik.de/English/Motors-Brushless.htm, (12/19/2009).

[11] XSENS.
http://www.xsens.com/, (12/19/2009).

[12] D. Anderson. Modification of a generalized inverse simulation technique for rotorcraft flight. *Proceedings of the Institution of Mechanical Engineers, Part G: Journal of Aerospace Engineering*, 217(2003):61–73, 2003.

[13] V. A. Anikin, B. A. Vassiliev, and V. N. Kvokov. Modeling of Coaxial Helicopter Flight Characteristics. In *27th European Rotorcraft Forum*, Moscow, Russia, 2001.

[14] O. A. Bauchau and J. Wang. Efficient and Robust Approaches for Rotorcraft Stability Analysis. In *American Helicopter Society 63rd Annual Forum*, Virginia Beach, USA, 2007.

[15] M. Benedict, T. Jarugumilli, and I. Chopra. Design and Development of a Hover-Capable Cyclocopter Micro Air Vehicle. In *American Helicopter Society 65th Annual Forum*, Grapevine, USA, 2009.

[16] C. Bermes, S. Leutenegger, S. Bouabdallah, D. Schafroth, and R. Siegwart. New Design of the Steering Mechanism for a Mini Coaxial Helicopter. In *IEEE International Conference on Intelligent Robots and Systems (IROS)*, Nice, France, 2008.

[17] C. Bermes, D. Schafroth, S. Bouabdallah, and R. Siegwart. Design Parameters for Coaxial Rotary Wing MAVs with Passive Roll and Pitch Stability. In *American Helicopter Society 65th Annual Forum*, Grapevine, USA, 2009.

[18] C. Bermes, D. Schafroth, S. Bouabdallah, and R. Siegwart. Modular Simulation Model for Coaxial Rotary Wing MAVs. In *2nd International Symposium on Unmanned Aerial Vehicles*, Reno, USA, 2009.

[19] E. Bernet. *Der RC Hubschrauber.* Verlag für Technik und Handwerk, Baden-Baden, 2006. (in German).

[20] S. Bhandari and R. Colgren. 6-DoF Dynamic Model for a Raptor 50 UAV Helicopter Including Stabilizer Bar Dynamics. In *AIAA Modeling and Simulation Technologies Conference and Exhibit*, Keystone, USA, 2006.

[21] W. Bittner. *Flugmechanik der Hubschrauber.* Springer, 2nd edition, 2005. (in German).

[22] F. Bohorquez and D. Pines. Hover Performance and Swashplate Design of a Coaxial Rotary Wing Micro Air Vehicle. In *American Helicopter Society 60th Annual Forum*, Baltimore, USA, 2004.

[23] F. Bohorquez, F. Rankinsy, J. Baederz, and D. Pines. Hover Performance of Rotor Blades at Low Reynolds Numbers for Rotary Wing Micro Air Vehicles. An Experimental and CFD Study. In *AIAA Applied Aerodynamics Conference*, Orlando, USA, 2003.

[24] S. Bouabdallah. *Design and Control of Quadrotors with Application to Autonomous Flying.* PhD thesis, EPF Lausanne, 2007.

[25] S. Bouabdallah, G. Caprari, and R. Siegwart. Design and Control of an Indoor Coaxial Helicopter. In *IEEE International Conference on Intelligent Robots and Systems (IROS)*, Beijing, China, 2006.

[26] S. Bouabdallah, P. Murrieri, and R. Siegwart. Towards Autonomous Indoor Micro VTOL. *Autonomous Robots*, 18:171–183, 2005.

[27] A. Bramwell. *Helicopter Dynamics*. Butterworth Heinemann, 2nd edition, 2001.

[28] N. Cartwright. PicooZ and Gyrotor. *Model Helicopter World*, (12):18–22, 2006.

[29] L. Chen and P. McKerrow. Modelling the Lama Coaxial Helicopter. In *Australasian Conference on Robotics & Automation*, Brisbane, Australia, 2007.

[30] R. T. N. Chen. Effects of Primary Rotor Parameters on Flapping Dynamics. Technical report, NASA, 1980.

[31] T. Cheviron, A. Chriette, and F. Plestan. Generic Nonlinear Model of Reduced Scale UAVs. In *IEEE International Conference on Robotics and Automation (ICRA)*, Kobe, Japan, 2009.

[32] C. P. Coleman. A Survey of Theoretical and Experimental Coaxial Rotor Aerodynamic Research. Technical report, NASA, 1997.

[33] D. Cooper, J. Cycon, and S. Moore. Sikorsky Aircraft Unmanned Aerial Vehicle (UAV) Program. In *19th European Rotorcraft Forum*, Cernobbio, Italy, 1993.

[34] R. Cunha. *Modeling and Control of an Autonomous Robotic Helicopter*. MS Thesis, Universidade Tecnica de Lisboa, 2002.

[35] R. Cunha. *Advanced Motion Control for Autonomous Air Vehicles*. PhD Thesis, Universidade Tecnica de Lisboa, 2007.

[36] J. P. Cycon. Decoding the Cypher UAV. *Vertiflite*, 36:56–58, 1990.

[37] H. Dresig and F. Holzweißig. *Maschinendynamik*. Springer, 7th edition, 2006. (in German).

[38] A. Dzul, T. Hamel, and R. Lozano. Modeling and Nonlinear Control for a Coaxial Helicopter. In *IEEE International Conference on Systems, Man and Cybernetics*, volume 6, 2002.

[39] D. Everson, P. Samuel, and D. Pines. Determination of Rotary Wing MAV Stability Derivatives in Hover using a Forced Oscillation Test Stand. In *American Helicopter Society 61st Annual Forum*, Grapevine, USA, 2005.

[40] P. Ferrat, C. Gimkiewicz, S. Neukom, Y. Zha, A. Brenzikofer, and T. Baechler. Ultraminiature omnidirectional camera for an autonomous flying micro-robot. In *Proceedings of SPIE*, volume 7000, Strasbourg, France, 2008.

[41] C. Friedman, A. Fertman, and O. Rand. A Generic Rotorcraft Simulation Using Matlab/Simulink. In *American Helicopter Society 65th Annual Forum*, Grapevine, USA, 2009.

[42] D. Fusato and R. Celi. Design Sensitivity Analysis for Helicopter Flight Dynamic and Aeromechanic Stability. *Journal of Guidance, Control, and Dynamics*, 26(6):918–927, 2003.

[43] D. Fusato and R. Celi. Multidisciplinary Design Optimization for Aeromechanics and Handling Qualities. *Journal of Aircraft*, 43(1):241–252, 2006.

[44] U. Gerber. *Die Entwicklung des sowjetischen Hubschrauberbaus*. Elbe-Dnjepr Verlag, 2002. (in German).

[45] K. Graichen, V. Hagenmeyer, and M. Zeitz. A new approach to inversion-based feedforward control design for nonlinear systems. *Automatica*, 42:2033–2041, 2005.

[46] K. Graichen and M. Zeitz. Feedforward Control Design for Finite-Time Transition Problems of Nonlinear Systems with Input and Output Constraints. *IEEE Transactions on Automatic Control*, 53(5):1273–1278, 2008.

[47] S. Grzonka, G. Grisetti, and W. Burgard. Towards a Navigation System for Autonomous Indoor Flying. In *IEEE International Conference on Robotics and Automation (ICRA)*, Kobe, Japan, 2009.

[48] C. Haosheng and C. Darong. Identification of a Model Helicopters Yaw Dynamics. *Journal of Dynamic Systems, Measurement, and Control*, 127(1):140–145, 2005.

[49] H. R. Harrison and T. Nettleton. *Advanced Engineering Dynamics*. Butterworth-Heinemann, 1997.

[50] B. Heimann, W. Gerth, and K. Popp. *Mechatronik*. Hanser, 2006. (in German).

[51] Y. Higashi, K. Tanaka, H. Ohtake, and H. O. Wang. Construction of Simulation Model of a Flying Robot with Variable Attack Angle Mechanism. In *IEEE International Conference on Intelligent Robots and Systems (IROS)*, San Diego, USA, 2007.

[52] W. Hirosue, A. Ookura, and S. Sunada. A study of a coaxial helicopter(ii): Analysis on effects of a stabilizer bar on fuselage motion. In *41st Aircraft Symposium of The Japan Society for Aeronautical and Space Sciences*, 2003. (in Japanese).

[53] M. Inglin. *From Kamov's first design to Micro Aerial Vehicles - a survey on coaxial helicopters*. Studies on Mechatronics, ETH Zürich, 2009.

[54] E. Johnson and M. Turbe. Modeling, Control, and Flight Testing of a Small Ducted-Fan Aircraft. *Journal of Guidance, Control, and Dynamics*, 29(4):769–779, 2006.

[55] M. Karpelson, G.-Y. Wei, and R. Wood. A Review of Actuation and Power Electronics Options for Flapping-Wing Robotic Insects. In *IEEE International Conference on Robotics and Automation (ICRA)*, Pasadena, USA, 2008.

[56] K. Kondak, C. Deeg, G. Hommel, M. Musial, and V. Remuss. Mechanical Model and Control of an Autonomous Small Size Helicopter with a Stiff Main Rotor. In *IEEE International Conference on Intelligent Robots and Systems (IROS)*, Sendai, Japan, 2004.

[57] N. Koratkar and I. Chopra. Development of a Mach-Scaled Model with Piezoelectric Bender Actuated Trailing-Edge Flaps for Helicopter Individual Blade Control (IBC). *AIAA Journal*, 38(7):1113–1124, 2000.

[58] R. Kube. *Einfluss der Blattelastizität auf die höherharmonische Steuerung und Regelung eines gelenklosen Hubschrauberrotors*. PhD Thesis, Technische Universität Braunschweig, 1997. (in German).

[59] J. Leishman. *Helicopter Aerodynamics*. Cambridge, 2nd edition, 2006.

[60] S. Leutenegger. *Mechanical Design and Realization of a Steering Mechanism for a Coaxial Helicopter*. Semester Thesis, ETH Zürich, 2007.

[61] S. Lindemann. *Model Updating an einem biegeelastischen Rotor*. PhD Thesis, University of Kassel, 2009. (in German).

[62] L. C. Mak et al. Design and development of the Micro Aerial Vehicles for Search, Tracking And Reconnaissance (MAVSTAR) for MAV08. In *1st US-Asian Demonstration and Assessment of Micro-aerial and Unmanned Ground Vehicle Technology (MAV08)*, Agra, India, 2008.

[63] B. Mettler. *Identification Modeling and Characteristics of Miniature Rotorcraft*. Kluwer Academic Publishers, 2003.

[64] B. Mettler, C. Dever, and E. Feron. Scaling Effects and Dynamic Characteristics of Miniature Rotorcraft. *Journal of Guidance, Control, and Dynamics*, 27(3):466–478, 2004.

[65] P. Murren. Rotor and Aircraft Passively Stable in Hover, Patent WO 2004/103814A1, 2004.

[66] P. Murren. Passively stable rotor system for indoor hovering UAS. In *International Powered Lift Conference*, London, UK, 2008.

[67] L. R. Newcome. *Unmanned Aviation: A Brief History of Unmanned Aerial Vehicles.* American Institute of Aeronautics and Astronautics, 2004.

[68] T. T. H. Ng and G. S. B. Leng. Application of genetic algorithms to conceptual design of a micro-air vehicle. *Engineering Applications of Artificial Intelligence*, 15:439–445, 2002.

[69] T. T. H. Ng and G. S. B. Leng. Design optimization of rotary-wing micro air vehicles. *Proceedings of the Institution of Mechanical Engineers, Part C: Journal of Mechanical Engineering Science*, 220(6):865–873, 2006.

[70] T. T. H. Ng and G. S. B. Leng. Design of small-scale quadrotor unmanned air vehicles using genetic algorithms. *Proceedings of the Institution of Mechanical Engineers, Part G: Journal of Aerospace Engineering*, 221(5):893–905, 2007.

[71] A. Oppenheim and W. Schafer. *Discrete-time signal processing.* Prentice Hall, Upper Saddle River (N.J.), 2nd edition, 1999.

[72] A. J. Pappas. Flight testing of drone helicopters. *Annals New York Academy of Sciences*, 107:25–39, 1963.

[73] J. L. Pereira, D. Bawek, and I. Chopra. Design and Development of a Quad-Shrouded-Rotor Micro Air Vehicle. In *American Helicopter Society 65th Annual Forum*, Grapevine, USA, 2009.

[74] M. Perhinschi and J. V. R. Prasad. A Simulation Model of an Autonomous Helicopter. In *13th Bristol International RPV/UAV Systems Conference*, Bristol, UK, 1998.

[75] P. Pounds and R. Mahony. Design Principles of Large Quadrotors for Practical Applications. In *IEEE International Conference on Robotics and Automation (ICRA)*, Kobe, Japan, 2009.

[76] P. Pounds, R. Mahony, and J. Gresham. Towards Dynamically-Favourable Quad-Rotor Aerial Robots. In *Australasian Conference on Robotics and Automation*, Canberra, Australia, 2004.

[77] R. W. Prouty. *Helicopter Performance, Stability, and Control.* PWS Engineering, Boston, 1986.

[78] S. Pulla and A. Lal. Insect Powered Micro Air Vehicles. In *IEEE International Conference on Robotics and Automation (ICRA)*, Kobe, Japan, 2009.

[79] O. Rand and V. Khromov. Helicopter Sizing by Statistics. *Journal of the American Helicopter Society*, 49(3):300–317, 2004.

[80] J. F. Roberts, J.-C. Zufferey, and D. Floreano. Energy Management for Indoor Hovering Robots. In *IEEE International Conference on Intelligent Robots and Systems (IROS)*, Nice, France, 2008.

[81] P. Samuel, J. Sirohi, F. Bohorquez, and R. Couch. Design and Testing of a Rotary Wing MAV with an Active Structure for Stability and Control. In *American Helicopter Society 61st Annual Forum*, Grapevine, USA, 2005.

[82] D. Schafroth. *Aerodynamics, Modeling and Control of an Autonomous Micro Helicopter*. PhD Thesis, ETH Zürich, 2010.

[83] D. Schafroth, C. Bermes, S. Bouabdallah, and R. Siegwart. Modeling and System Identification of the muFly Micro Helicopter. *Journal of Intelligent and Robotic Systems*, 57(1-4):27–47, 2009.

[84] D. Schafroth, S. Bouabdallah, C. Bermes, and R. Siegwart. From the Test Benches to the First Prototype of the muFly Micro Helicopter. *Journal of Intelligent and Robotic Systems*, 54(1-3):245–260, 2008.

[85] D. Schlüter. *Die Geschichte des Modellhubschraubers und andere Erinnerungen*. Neckar-Verlag, 2007. (in German).

[86] D. Schlüter. *Hubschrauber ferngesteuert*. Neckar-Verlag, 2007. (in German).

[87] I. A. Simons and A. N. Modha. Technical Note Gyroscopic Feathering Moments and the 'Bell Stabilizer Bar' on Helicopter Rotors. *Journal of the American Helicopter Society*, 52(1), 2007.

[88] S. Sunada, H. Sumino, A. Matsue, and H. Tokutake. Analysis of Flapping Motion of Rotors of Small Coaxial Helicopter. *Transactions of the Japan Society for Aeronautical and Space Sciences*, 49(164):101–108, 2006.

[89] N. Tsuzuki, S. Sato, and T. Abe. Design Guidelines of Rotary Wings in Hover for Insect-Scale Micro Air Vehicle Applications. *Journal of Aircraft*, 44:252–263, 2007.

[90] E. R. Ulrich, D. J. Pines, and S. Gerardi. Autonomous Flight of a Samara MAV. In *American Helicopter Society 65th Annual Forum*, Grapevine, USA, 2009.

[91] C. M. Velez, A. Agudelo, and J. Alvarez. Modeling, Simulation and Rapid Prototyping of an Unmanned Mini-Helicopter. In *AIAA Modeling and Simulation Technologies Conference*, Keystone, USA, 2006.

[92] J. A. Vilchis, B. Brogliato, A. Dzul, and R. Lozano. Nonlinear modelling and control of helicopters. *Automatica*, 39:1583–1596, 2003.

[93] H. Wang, D. Wang, and X. Niu. Modeling and Hover Control of a Novel Unmanned Coaxial Rotor/Ducted-Fan Helicopter. In *IEEE International Conference on Automation and Logistics*, Jinan, China, 2007.

[94] W. Wang, G. Song, K. Nonami, M. Hirata, and O. Miyazawa. Autonomous Control for Micro-Flying Robot and Small Wireless Helicopter X.R.B. In *IEEE International Conference on Intelligent Robots and Systems (IROS)*, Beijing, China, 2006.

[95] J. Watkinson. *The Art of the Helicopter*. Elsevier, 2004.

[96] L. A. Young, J. L. Johnson, R. Demblewski, J. Andrews, and J. Klem. New Concepts and Perspectives on Micro-Rotorcraft and Small Autonomous Rotary-Wing Vehicles. In *20th AIAA Applied Aerodynamics Conference*, St. Louis, USA, 2002.

[97] H. Youngren, S. Jameson, and B. Satterfield. Design of the Samarai Monowing Rotorcraft Nano Air Vehicle. In *American Helicopter Society 65th Annual Forum*, Grapevine, USA, 2009.

[98] R. Zimmermann. *Rotor Characterization for muFly*. Semester Thesis, ETH Zürich, 2008.

i want morebooks!

Buy your books fast and straightforward online - at one of world's fastest growing online book stores! Environmentally sound due to Print-on-Demand technologies.

Buy your books online at
www.get-morebooks.com

Kaufen Sie Ihre Bücher schnell und unkompliziert online – auf einer der am schnellsten wachsenden Buchhandelsplattformen weltweit! Dank Print-On-Demand umwelt- und ressourcenschonend produziert.

Bücher schneller online kaufen
www.morebooks.de

 VDM Verlagsservicegesellschaft mbH
Heinrich-Böcking-Str. 6-8   Telefon: +49 681 3720 174   info@vdm-vsg.de
D - 66121 Saarbrücken   Telefax: +49 681 3720 1749   www.vdm-vsg.de

Printed by Books on Demand GmbH, Norderstedt / Germany